Sustainable Development Goals Series

The **Sustainable Development Goals Series** is Springer Nature's inaugural cross-imprint book series that addresses and supports the United Nations' seventeen Sustainable Development Goals. The series fosters comprehensive research focused on these global targets and endeavours to address some of society's greatest grand challenges. The SDGs are inherently multidisciplinary, and they bring people working across different fields together and working towards a common goal. In this spirit, the Sustainable Development Goals series is the first at Springer Nature to publish books under both the Springer and Palgrave Macmillan imprints, bringing the strengths of our imprints together.

The Sustainable Development Goals Series is organized into eighteen subseries: one subseries based around each of the seventeen respective Sustainable Development Goals, and an eighteenth subseries, "Connecting the Goals", which serves as a home for volumes addressing multiple goals or studying the SDGs as a whole. Each subseries is guided by an expert Subseries Advisor with years or decades of experience studying and addressing core components of their respective Goal.

The SDG Series has a remit as broad as the SDGs themselves, and contributions are welcome from scientists, academics, policymakers, and researchers working in fields related to any of the seventeen goals. If you are interested in contributing a monograph or curated volume to the series, please contact the Publishers: Zachary Romano [Springer; zachary.romano@springer.com] and Rachael Ballard [Palgrave Macmillan; rachael.ballard@palgrave.com].

Deny Hidayati · Yanu Endar Prasetyo ·
Jonatan Lassa · Syarifah Aini Dalimunthe
Editors

Examining Disaster Risk Reduction in Indonesia

Building Social Resilience

Springer

Editors
Deny Hidayati
National Research and Innovation Agency
Jakarta, Indonesia

Yanu Endar Prasetyo
National Research and Innovation Agency
Jakarta, Indonesia

Jonatan Lassa
Geology and Nuclear Sciences (GNS Science)
Avalon, Wellington, New Zealand

Syarifah Aini Dalimunthe
National Research and Innovation Agency
Jakarta, Indonesia

ISSN 2523-3084 ISSN 2523-3092 (electronic)
Sustainable Development Goals Series
ISBN 978-981-96-4181-9 ISBN 978-981-96-4182-6 (eBook)
https://doi.org/10.1007/978-981-96-4182-6

This Springer imprint is published by the registered company Springer Nature Singapore Pte Ltd.
The registered company address is: 152 Beach Road, #21-01/04 Gateway East, Singapore 189721, Singapore

Foreword

We live as part of a society where we receive education, work, earn a living, and spend time with our families. However, disasters can strike unexpectedly, posing a constant threat to the life and property we have built. Since we cannot predict when a disaster will occur, developing systems that reduce disaster risk and minimize potential damage through proactive disaster risk reduction (DRR) activities is essential.

What is "disaster risk reduction (DRR)"? To briefly answer this question, let me return to my personal memory. The term might have existed for more than forty years. When I was in kindergarten, my mother prepared a cushion for me to sit in a chair and bring it to kindergarten. These cushions were used not only to sit on the chair comfortably but also to protect my head in the event of an earthquake. I attended earthquake evacuation drills in which cushions were utilized at many kindergartens. In elementary school, I participated in community activities, evacuation drills, and community cleanups in our town. DRR activities had taken root as community activities and were not something special. In order to protect our families and property, DRR needs to be an integral part of our daily lives.

This book provides a comprehensive exploration of Indonesia's multifaceted challenges and responses associated with disaster risk and resilience. Each study within this collection examines a critical dimension of how communities, institutions, and governing bodies navigate the complex landscapes of crises—from pandemics to natural hazard-induced disasters. This book paints a holistic picture of resilience-building in one of the world's most disaster-prone regions by weaving together insights on social capital, risk governance, mental health, and cultural considerations.

The included works span a broad array of themes. They highlight the importance of recognizing local contexts and culture in shaping effective risk communication and governance, the pivotal role of community resilience and social capital in adapting to widespread disruptions, and the necessity of developing sustainable policy frameworks—such as adaptive social protection—to ensure inclusive post-disaster recovery. Attention to mental health and psychosocial support, as well as the feasibility of financial mechanisms like flood insurance, further underscores the complexity of strengthening preparedness and reducing vulnerability. By examining efforts from diverse vantage points—media representations, civil society's role, foreign policy implications, and the integration of digital solutions for marginalized groups—this book not only catalogues the rich tapestry of disaster risk reduction initiatives but also offers a forward-looking perspective.

Indonesia is as disaster-prone as Japan is. However, it is necessary to learn from past disasters and take effective disaster countermeasures to reduce risks before disasters and to quickly and appropriately rebuild and recover after disasters in both countries. Successful disaster management and resilience-building hinge not on a singular approach but on harmonizing multiple strategies, acknowledging local realities, and fostering collaborative governance structures. As such, this book aims to inform policymakers, practitioners, and stakeholders seeking to enhance Indonesia's resilience in an era of growing uncertainty and interconnected risk.

Makoto Ikeda (Dr./Mr.)
Senior Researcher, Asian Disaster
Reduction Center (ADRC)
Visiting Associate Professor
Research Center for Urban Safety
and Security (RCUSS)
Kobe University
Kobe, Japan

Preface

In today's increasingly complex and unpredictable world, disaster risk reduction demands a nuanced understanding of both the physical and social dimensions of risk. This edited volume, "*Examining Disaster Risk Reduction in Indonesia*," brings together a collection of chapters that explore how Indonesia, as one of the world's most disaster-prone nations, navigates the complexities of disaster management through the lens of social science. Each chapter builds upon the foundational idea that disaster risk is not merely a technical/technological or natural challenge. Disaster risk and disaster risk reduction are deeply embedded in the cultural, social, and economic fabric of communities.

This book consists of 11 chapters. The first chapter, "*The Role of Social Capital in Enhancing the Community's Adaptation to Pandemic Disruption in Indonesia*," emphasizes the vital role of social capital during the COVID-19 pandemic. Drawing on empirical evidence from various regions, the authors highlight how community networks and local wisdom became crucial in supporting vulnerable populations when formal systems were overwhelmed. Initiatives like *Jogo Tonggo* in Central Java and *umplungan* in East Java show how communities quickly self-organized and/or mobilized to provide food, healthcare access, and emotional support, leveraging trust and pre-existing social ties. Digital platforms, such as WhatsApp, also played an important role in maintaining these networks and facilitating community-led relief efforts despite mobility restrictions. This chapter underscores that social capital is essential for immediate crisis response and building long-term resilience. Communities with strong social bonds are better equipped to navigate future challenges, whether they stem from pandemics, natural hazard-induced disasters, or economic disruptions. By integrating local knowledge and community support into disaster management strategies, the authors argue that social cohesion is crucial for effective and sustainable disaster risk reduction.

The second chapter, "*Towards the Development of an Ideal Risk Communication by Acknowledging Local Culture*," focuses on the need to incorporate local knowledge and cultural values into risk communication strategies. It examines the traditional top-down approach, often overlooking local communities' unique perceptions and practices. The authors argue that disaster warnings and mitigation efforts are far more effective when they respect and integrate local worldviews, rather than imposing standardized messages that may not resonate with local context. The chapter highlights the importance of understanding cultural contexts in disaster management by advocating for a more flexible, two-way dialogue between the government and local communities. To be truly effective, risk communication must not only disseminate scientific knowledge but also build on the lived experiences and cultural beliefs of the communities it aims to protect. This approach fosters trust and ensures disaster preparedness strategies are more locally relevant and accepted.

The third chapter, "*Considering Risk Culture to Strengthen Disaster Risk Governance in Indonesia*," expands this discussion by focusing on the concept of 'risk culture.' The authors stress that successful disaster risk governance cannot ignore the local belief systems and values influencing how communities understand and respond to risk. These cultural factors shape people's attitudes toward disaster preparedness and willingness to engage with governance efforts. By incorporating risk culture into governance strategies, disaster management becomes more aligned with the community's worldview, making it easier to implement and more effective in building long-term resilience.

Building on the importance of understanding local belief systems and cultural values in disaster governance, the next chapter focuses on the equally critical dimension of mental and emotional well-being. While cultural and social factors shape how communities prepare for and respond to disasters, the psychological aftermath can be just as profound. The fourth chapter, "*Unravelling Mental Health Issues and Psychosocial Support Post-disaster in Central Sulawesi*," underscores this by highlighting the emotional and psychological toll disasters can take. It emphasizes that effective disaster management must address not only physical reconstruction but also the mental health of affected populations, advocating for psychosocial support as an essential component of long-term recovery.

Shifting from the personal and emotional impacts of disasters, the discussion moves toward the broader societal influences on disaster preparedness. Beyond individual and community responses, how risks are communicated and perceived on a larger scale plays a crucial role in shaping preparedness efforts. The fifth chapter, "*Framing Disaster Preparedness: Media Perspectives*," delves into the significant role media plays in influencing public understanding of risk. The authors explore how media narratives can shape societal attitudes toward disaster preparedness, pointing out both the strengths and limitations of media as a tool for educating the public. The chapter underscores the importance of accurate, responsible reporting to raise awareness and encourage proactive responses to potential disasters.

Following exploring the media's role in shaping disaster preparedness, the volume transitions to more pragmatic approaches to disaster risk reduction. While communication and social factors are vital, practical financial mechanisms can also play a crucial role in helping communities recover from disasters. In the sixth chapter, "*Is Flood Insurance Scheme in Regional Areas Feasible? An Explorative Study in Riau Indonesia*," the authors investigate the potential of flood insurance as a tool to mitigate financial vulnerability in one of Indonesia's most flood-prone regions. This chapter provides a fresh perspective on how economic strategies, such as insurance schemes, can help reduce the long-term financial impacts of disasters, offering valuable insights into the practical intersection of finance and disaster management.

Expanding the scope to international dimensions, the seventh chapter, "*Indonesia's Foreign Policy on the Status of Forces Agreement (SOFA) in Post Natural Hazard Induced-Disasters Contexts*," explores how Indonesia engages with foreign military assistance in post-disaster recovery efforts. This chapter highlights the complexities of international collaboration, shedding light on the diplomatic, logistical, and sovereignty issues that arise when foreign forces are involved in disaster response efforts. It provides important insights into how international partnerships can enhance, or at times complicate, recovery in the wake of significant natural hazard-induced disasters.

The final chapters bring the focus back to resilience and social protection. "*The Role of Community Resilience in Disaster Risk Reduction*" emphasizes the strength of community-based resilience initiatives, arguing that local communities are often the first and most effective responders in the face of disaster. The chapter "*Adaptive Social Protection (ASP) for Post-disaster Recovery in Indonesia: Building Resilience and Inclusion through Intersectional Framework*" synthesizes the book's broader themes, advocating for an intersectional approach to social protection that addresses not only immediate disaster recovery but also long-term resilience.

Reflecting on the early stages of the COVID-19 pandemic, the closing chapter, "*Fixing the State? A Reflection on Civil Society Response to COVID-19 Crisis in Indonesia*", examines the behavioral traits of CSOs and NGOs as policy entrepreneurs who sought to 'fix the state' and 'jump-start local governance' during crises and disasters. The COVID-19 crisis provided a unique opportunity to observe the distinctive behavioral traits shared by Indonesian CSOs.

Examining Disaster Risk Reduction in Indonesia offers a comprehensive exploration of disaster risk in Indonesia, weaving together cultural, social, economic, and policy perspectives. Each chapter contributes to a holistic understanding of how disaster management must be approached with an awareness of the intricate social structures and cultural values that shape human response to risk. I hope this volume will be a valuable resource for scholars, policymakers, and practitioners seeking to develop more effective and socially informed disaster risk reduction strategies.

Jakarta, Indonesia — Deny Hidayati
Jakarta, Indonesia — Yanu Endar Prasetyo
Avalon, New Zealand — Jonatan Lassa
Jakarta, Indonesia — Syarifah Aini Dalimunthe

Contents

The Role of Social Capital in Enhancing the Community's Adaptation to Pandemic Disruption in Indonesia

1

Deny Hidayati and Wida Widayatun

Abstract

The COVID-19 pandemic threatens life safety and has a wider impact, disrupting community social life and livelihoods in many countries, including Indonesia. The pandemic 'forces' the community to limit their mobility and socio-economic activities, thereby hampering health services, livelihoods, meeting basic needs, socio-cultural, and religious activities. This article aims to discuss the important role of the social capital of a community in increasing its capacity to deal with pandemic disruption and adapt to new normal conditions. This article is based on empirical evidence from community experiences in dealing with the pandemic in various regions of Indonesia, especially in areas that were the epicentre of the spread of the coronavirus. Data was collected through research using qualitative methods with open interview techniques. The community played an important role in supporting its members by applying coping strategies based on local knowledge and wisdom practiced for a long time in their areas. Community members helped each other to meet the needs for food, basic foodstuffs, and other basic needs when COVID-19 patients and their family members were self-isolating and maintained their health conditions during and after the pandemic. Community support was also carried out by distributing information via community digital communication networks, especially WhatsApp groups. This article emphasizes the importance of maintaining community social capital and accelerating the transformation of virtual communication through social networks to increase community digital communication literacy and access to communication technology to adapt to disruption and anticipate future pandemics.

Keywords

Social capital · Community's adaptation · Pandemic disruption · Indonesia

D. Hidayati · W. Widayatun (✉)
Research Center for Population - The National Research and Innovation Agency (BRIN), Jakarta, Indonesia
e-mail: wida001@brin.go.id

1.1 Introduction

The COVID-19 pandemic is a health disaster in many countries, including Indonesia. The pandemic threatens life safety and disrupts people's lives and livelihoods. Older people were encouraged to stay away from crowds and stay home to avoid the coronavirus. Even though most Indonesian people have the opposite habit,

D. Hidayati et al. (eds.), *Examining Disaster Risk Reduction in Indonesia*,
Sustainable Development Goals Series, https://doi.org/10.1007/978-981-96-4182-6_1

they always practice direct contact when meeting friends or relatives, shake hands, and hug each other. The implementation of physical and social distancing restrictions has "forced" them to limit their mobility and socio-economic activities. This condition creates many obstacles, especially access to health facilities and services, livelihoods, meeting basic needs, and daily activities. Socialization about the pandemic from government authorities was initially limited. At the same time, information sourced from social media was quite massive even though it was not necessarily true, so there was often confusion of information and hoaxes. This condition causes panic and social anxiety and increases the socio-economic vulnerability of society (Hidayati et al. 2020; Ngadi et al. 2020; Saifudin 2020).

COVID-19 targets all age groups, with a vast number of positive coronavirus cases. In Indonesia, confirmed cases of COVID-19 reached 6,813,429 with 161,918 deaths since it was announced on March 2, 2020, to November 6, 2023 (COVID-19.go.id 2023). The deaths of COVID-19 patients are mainly caused by health factors, especially those who have symptoms and comorbidities (Michelen et al. 2020; Promislow and DPhil1 2020; Shahid et al. 2020; WHO 2020; Liu et al. 2020). Death cases are dominated by older people aged 60 years and over, especially COVID-19 patients who have degenerative diseases. The elderly have more illnesses than young people (Doloksaribu and Verayanti Manalu 2021; Sari and Daryanto 2021). The body's immunity decreases with increasing age, so the risk of older people contracting the coronavirus is greater and increases (Al Mubarroh et al. 2021).

Social Restriction Policies during the pandemic affect psychosocial development (Marliana et al. 2022), simultaneously disrupt lifestyles and cause social isolation, thereby increasing emotional stress (Reger et al. 2020). The pandemic causes excessive anxiety and worry that the mind cannot control (Chigwedere et al. 2021; Okuyama et al. 2021). This disaster is a life event associated with uncertainty, ambiguity, and loss of control, which triggers stress in a person (Shanahan et al. 2022; Reger et al. 2020). This symptom can be experienced by anyone, characterized by physical changes in anxiety, such as excessive sweating, panic, depression, chronic stress, palpitations, anxiety, and difficulty breathing. Individuals who experience this kind of anxiety usually cannot describe what kind of anxiety it is to others (Evans et al. 2018; Stansfeld and Candy 2006; Xiong et al. 2020).

The community has made various efforts to deal with the impact of the pandemic. They establish social bonds (Gavriluță et al. 2022a, b; Pabbajah et al. 2022), which assist a person develop a more positive attitude to become more resilient. They also increase their capacity through intellectual formation, such as seeking information about the pandemic from official sources so that they are not influenced by fake news circulating in the mass media and the surrounding environment. In addition, they use a spiritual approach or increase religiosity (Coppola et al. 2021; Moroń et al. 2021; Moghadam et al. 2021) by praying, reading the Qur'an, *sholawat* and hizib to improve well-being in conditions which is unstable during a pandemic (Baidowi et al. 2021).

1.2 Literature Review

1.2.1 Pandemic

The COVID-19 pandemic has presented challenges to communities around the world, causing widespread morbidity and mortality (Wang et al. 2022; Brown et al. 2022; World Health Organization 2020). The COVID-19 pandemic has also had a significant impact on people's lives, not only in medical aspects, but also in various other aspects such as the economy (Park and Cho 2022; Ilukor et al. 2022; Zhang et al. 2021), sociological (Bhardwaj 2021; Pascaru 2021), religious practices (Coppola et al. 2021; Moroń et al. 2021; Moghadam et al. 2021) to psychological aspects (Marchi et al. 2022; Abdullah 2020; Aguilar-Latorre et al. 2022). Bidzan-Bluma et al. (2020) stated that quality of life, life satisfaction, and well-being during the pandemic are influenced by age. Regarding psychological aspects, older people are one of the most affected groups in society. Social distancing policies during the pandemic increase loneliness and hinder psychosocial development (Marliana et al. 2022). COVID-19 has also increased levels of isolation in young people and impacted their ability to draw on the support of social networks (Ford et al. 2021). The particularly detrimental impact that enforced lockdowns and school closures have had on young people's mental health (Loades et al. 2022). The COVID-19 pandemic also presents various kinds of negative stigma for society. The impact of this stigma can have a huge impact on the mental and psychological health of sufferers (Elgar et al. 2020; Hadi 2020).

1.2.2 Social Capital of Community

The COVID-19 pandemic has had many negative impacts on people's lives and efforts are needed to minimize the impact and return life to normal. Therefore, social capital needs to be presented comprehensively to mitigate the effects of the COVID-19 pandemic (Borgonovi and Andrieu 2020; Borgonovi et al. 2021; Cowan 2020). Social capital is believed to be one of the most potent formulations to deal with the impact of the COVID-19 pandemic, apart from other capital, such as economic capital and cultural capital. Social capital is the ability to carry out collective activities (working together) to achieve shared goals by an organization or group. Social capital is the ability to work together that is obtained from trust in a community group. If people trust each other, work together, and are based on existing universal values, there will be collective strength to help each other overcome problems or gaps resulting from pandemic disruption (Salman et al. 2021; Horwitz and Lascar 2021; Xiao et al. 2020).

Various forms of social capital, such as the advantages possessed by individuals or community groups, can be utilized or implemented to deal with the impact of the pandemic. For example, individuals or groups who have better skills or economic abilities can help other communities in need (Nicholas Pitas and Ehmer 2020). On the other hand, religious leaders can help through religious teachings in providing positive advice to other people during the pandemic. Likewise, psychiatrists, families, and other community groups can provide encouragement and motivational support for COVID-19 survivors (Hatabu et al. 2021). In this case, various elements or groups of society essentially need to work together to help others during the pandemic. Community knowledge and understanding regarding the role of social capital from cognitive, structural, and cognitive levels with commitment to implementation can create collaboration patterns between community groups and stakeholders to form effective institutions in efforts to handle the negative impacts of the pandemic (Alshurafat et al. 2021).

The pandemic has revealed the importance of humanity and confirmed that everything is interconnected. Digitally mediated communication tools play a greater role in generating social capital (Ellison et al. 2007). Digital and other forms of mediated communication have been utilized as a means of creating and fostering social relationships. Platforms like Facebook, Twitter, and Instagram are useful for disseminating factual, curated information about the pandemic and dispelling potentially harmful misinformation. Although social media has been linked to social capital in these contexts, existing research lacks discussion of this relationship (Pitas and Ehmer 2020).

1.2.3 Coping Strategies in Dealing with the Pandemic

Coping is the ability to respond and recover from something that causes stress. Coping is an effort to reduce the stress that occurs (Sarafino et al. 2015) or negative or positive behaviour and thoughts that can reduce the burden so as not to cause stress (Haber and Runyon 1984). This concept involves cognitive and behavioural strategies that are influenced by biological and psychological factors, including physical health, spirituality, and social support (Finlay et al. 2021).

A coping strategy is an effort to reduce pressure or threats to a person or group of people, or a community with the resources they have to solve problems. The concept of coping strategy varies according to references and experts. "Coping is a capacity to respond and to recover from something stressful" (Lazarus and Folkman 1984; Carver et al. 1989). "Coping strategies are behaviours that assist us in functioning better in certain situations or constantly changing cognitive and behavioural efforts to manage certain external and/or internal demands that are judged to be taxing" or "exceeding human resources" (Lazarus and Folkman 1984; Carver et al. 1989). This concept is a strategic process that includes feelings, thoughts, and actions (emotion, behaviour, or action). The strategy can be carried out by confronting and seeking social support (Lazarus and Folkman 1984).

Stuart and Sundeen (1991) pointed out two types of coping mechanisms: problem and emotion-focused coping. Problem-centred coping mechanisms are carried out in three ways: (1) confrontation by changing the situation or solving the problem aggressively, (2) isolation by withdrawing from the environment or not caring about the problem at hand, and (3) compromise by changing the situation with the heart. Meanwhile, mechanisms centred on emotions are carried out in various ways, including (1) denial of problems that occur, (2) rationalization with acceptable reasons, (3) compensation by covering up one's incompetence, (4) repression by forgetting unpleasant things, (5) projection by blaming others, etc.

Coping strategies were initially widely used in psychology to reduce psychological pressure or stress experienced by a person or certain group of people/communities, but recently, this concept has been used with a broader understanding and scope. WHO/EHA (1999) states, "coping strategies are cultural". This concept is based on the use of non-standard strategies adapted to the culture and needs of each individual or community group.

Research related to COVID-19 focuses on health aspects, such as maintaining the physical and mental health of the elderly during the pandemic, the risk of death from COVID-19, social distancing to avoid COVID-19 infection (Bavel et al. 2020; Razai et al. 2020). In fact, according to the World Health Organization (2008), health conditions are not only determined by physical and mental health, but also by a balance between physical, mental, and social functions. Research related to social support, especially socio-cultural practices, and the use of digital-based communication in facing and adapting during and after the pandemic using a qualitative approach is therefore still very much needed as learning material and developing concepts of community preparedness and adaptation in anticipating future pandemics.

This article generally aims to discuss the important role of the social capital of a community in increasing its capacity to deal with pandemic disruption and adapt to new normal conditions. The specific objectives include: (1) analyzing the community's understanding of the pandemic, (2) discussing the community's socio-cultural and religious practices when responding to pandemic disruption and adapting in efforts to restore post-pandemic disruption, and (3) discussing the use of digital-based communication through community social networks.

1.3 Method

1.3.1 Study Locations

This study was carried out in various regions in Indonesia, especially in the epicentre of the spread of COVID-19, including the Jabodetabek area, Greater Semarang and Greater Solo (Central Java Province), Greater Malang (East Java Province), and Bali Province. In each region, one or two districts/cities were selected as study locations. In the Jabodetabek area, Bekasi City was chosen because it is close to the centre of government, culture, and economy, thus attracting the interest of immigrants from various ethnic groups throughout Indonesia to work and live in this area. Semarang City and Wonogiri Regency in Central Java Province were locations that represent urban and rural locations with a high number of vulnerable residents (elderly) to the impact of the pandemic. Malang Regency in East Java Province was a location with a heterogeneous socio-cultural background of the community. Tabanan Regency in Bali Province was a location which describes the homogeneous socio-cultural conditions of the community. In each city and regency, one or two urban villages/villages were selected as data collection locations (Table 1.1).

Table 1.1 Study locations and characteristics of region

City/regency	Urban village/ village	Characteristics of urban village/ village
Bekasi city	Aren Jaya urban village	Urban area, multi-ethnics (Javanese, Sundanese, Batak, etc.), heterogeneous socio-cultural and economic of the community
Semarang city	Tanjung Mas urban village	Urban areas, heterogeneous people's livelihoods
	Sumurejo urban village	
Wonogiri regency	Manjung village	Rural areas, community homogeneity (Javanese) and livelihoods (agriculture)
	Kaliancar village	
Malang regency	Banjararum village	Areas with mixed characteristics between urban (residential) and rural (village) residents
	Sekarpuro village	
Tabanan regency	Dolot Peken village	Urban areas, heterogeneous people's livelihoods
	Kukuh village	Rural areas, homogeneous people's livelihoods (agriculture)

1.3.2 Data Collection

The research used qualitative methods with open interview techniques. The qualitative design aims to explore people's perceptions, meanings, interpretations, and experiences in dealing with the COVID-19 pandemic, as well as the efforts made by community members around where they live. Apart from that, the use of this method also aims to gain a deeper understanding of people's socio-cultural practices and the use of digital communication in responding to disruption so that they can restore their daily lives after the pandemic.

Open interviews were conducted based on the interview guides with the focus on exploring (1) the community's understanding of the COVID-19 pandemic, (2) the influence of the pandemic on the health and social and religious life of the community, (3) the obstacles they experience when accessing health services and carrying out social, religious activities, and using virtual communication to respond to the pandemic and its impact, and (4) the efforts they make in overcoming these obstacles.

Open interviews were conducted with community groups, including older people, the most vulnerable to being affected by the COVID-19 pandemic, caretakers, and selected members of the community who live in one housing complex/settlement, neighbourhood association/ community association (RT/RW) area, and village area. The participation of participants was completely voluntary; any information they provided would be kept confidential and quotes from their answers used only their first initials. They also had the right to stop participating at any time or refused to answer any questions. Apart from that, interviews were also conducted with resource persons at the community level, such as community leaders, traditional-cultural leaders, religious leaders, health cadres, *dasawisma,* and the COVID-19 Task Force as well as formal leaders at the village, sub-district, and district levels. They were selected purposively to explore more specific and interesting cases.

To gain a more complete understanding, researchers also conducted observations of the community when visiting the residences of community members and their families, as well as the environment where they carry out their daily activities. Data is also collected from secondary sources, including related documents, reports, research results, journal articles, and books.

1.3.3 Data Analysis

Qualitative data analysis applies thematic descriptive analysis methods. The analysis basically began when the researcher prepared interview guidelines, collected data at the research locations, and selected community caretakers and members who were interviewed. They were considered capable of providing complete and comprehensive information regarding the role of social capital in dealing with COVID-19 and adapting to pandemic disruption, especially through community socio-cultural practices and the use of virtual communication in responding to disruption due to the implementation of activity restrictions and health protocols during and after the pandemic. The results of open interviews were transcribed. The data is then sorted and grouped according to research themes. After that, synthesis and conclusions were carried out and linking the research findings with theories and results of previous similar research.

1.4 Results

1.4.1 Community Understanding of the Pandemic

Knowledge about the COVID-19 pandemic, followed by the implementation of health protocols clean and healthy living behaviour is important to prevent transmission of COVID-19. At the beginning of the pandemic, community knowledge of COVID-19 was still lacking. They did not know what the coronavirus was, how the virus works, and how it was transmitted. Information obtained from various sources was still very limited, and people were confused in finding correct information about COVID-19.

> I didn't know what the coronavirus was; the information was confusing and different among the sources of information. I didn't understand and to whom I should ask. They said that can be transmitted through the air…many people sick… the people here said that there was Pageblug. Mrs. G (Female, 60 years)

Public understanding of the crisis improved in line with the increasing dissemination of official information about COVID-19 from the national and local governments and stakeholders. The community obtains information about COVID-19 from various sources, namely social media such as YouTube, Facebook, and WhatsApp groups (family and community) and other media such as TV, Radio, and newspapers. Apart from that, information was also obtained through leaflets, brochures, and outreach materials from the COVID-19 Task Force and related stakeholders in each region. People know the symptoms, ways of transmission, and how to avoid contracting COVID-19. A good understanding of COVID-19 was also followed by discipline in implementing social distancing policies by most of the community.

Community knowledge and awareness of COVID-19 vaccination was low at the beginning of the vaccination programme. They were worried about side effects, and some still doubted the "halalness" (*kehalalan*) of the vaccine. After socialization about the importance of vaccination was carried out more intensively and was followed by increased access to services (including the pick-up and drop-off system), community awareness of vaccination increased.

1.4.2 Socio-cultural and Religious Practices in Responding and Adapting to Disruption

In facing the COVID-19 pandemic, the community applied local wisdom that has been passed down from generation to generation. They help each other and work together to help a member of the community who has been exposed to COVID-19. The community also used various traditional herbs to maintain their health condition. Apart from that, people are also increasing their religious activities to reduce fear and anxiety due to the COVID-19 pandemic.

Togetherness and mutual assistance in dealing with COVID-19 occurred in all study locations. In Central Java, efforts to help each other and protect community members facing COVID-19 are known as the local term "*Jogo Tonggo*" which is taken from the Javanese language. "Jogo" means guarding, while "*Tonggo*" means neighbour. Communities help each other and provide support to community members affected by COVID-19. These include sympathy and emotional support, providing food needs for all infected families during self-isolation and facilitating access to health services. When a family member in the neighbourhood was infected with COVID-19, other community members sent sympathy messages via WhatsApp or telephone. These messages were useful for maintaining the mental health of infected family members and also providing support for their families in carrying out treatment.

> Our neighbours sent messages and called while we were self-isolating. They provided support... encouragement and information about the various medicines and potions that need to be taken...very helpful and encouraging for us to recover from COVID-19. Mr H (Male, 71 years)

In helping each other, the community also provided food and nutritional needs for families who were exposed to COVID-19 and undergoing self-isolation. The community, through a neighbour (RT) caretaker, collects assistance (money or goods) from community members while still implementing social distancing. In East Java, collecting funds to help others is known as "*umplungan*". Donated money or goods were put into a container called "*umplung*" and managed to help community members affected by COVID-19.

> We visited resident's homes and collected donations from them. Residents put donated money into the "umplung", and the amount depends on the abilities of each resident. We manage this money to help residents who are self-isolating. Mrs A (Female, 59 years)

Apart from basic food and/or food assistance from neighbourhood caretakers, families who were self-isolating also received food assistance from local caring neighbours. The local caring neighbours (donors) voluntarily took turns providing food, vegetables, fruit, vitamins, and other necessities needed by a family who were self-isolating. In order to be useful assistance and meet the needs of the families who were self-isolating, the donors communicated the type of food needed.

Community members were also helping families who experienced symptoms of COVID-19 infection to access health services for prevention and treatment. This especially happens in elderly families and those living alone. The assistance provided by the community includes assisting consultations with health workers, facilitating rapid/antigen COVID-19 tests, and providing transportation to access health facilities.

> I had symptoms of fever and cough. I was scared to be infected with COVID-19 because I live alone. Neighbours provided assistance in carrying out COVID-19 tests and providing necessary medicines Mrs S (Female, 63 years).

People at the research location also consume traditional herbal drinks such as turmeric, ginger, lemongrass, and other herbal medicines to manage body immunity. The habit of drinking this traditional concoction has been passed down from generation to generation before the pandemic. This concoction was purchased from herbal medicine traders, online herbal shops, or they made it themselves. The habit of consuming traditional herbal supplements tends to increase during the pandemic.

Another effort made by the community to improve health conditions and reduce feelings of anxiety and fear due to COVID-19 was to increase religious activities. Most people carried out religious activities at home due to social restrictions, which have resulted in places of worship (mosques, churches, and temples) being closed in several places. Efforts were made to increase religious rituals per the religion adhered to by each member of society. Those who are Muslim, apart from praying 5 times a day, increase their "*dhikir*" and other "*sunnah*" worship. Followers of other religions also did the same thing according to their respective teachings.

The social distancing policy during the COVID-19 pandemic has given rise to new habits among people in carrying out religious rituals. They held recitations, "*tahlilan*", recitation of the Quran, Hindu religious celebrations (*e.g. Waisak*) and joint prayers for Christians using virtual meeting applications such as video calls, WhatsApp calls, and some via Google and Zoom meeting applications. Religious rituals online are a medium for meeting to maintain social relations between fellow religious communities without meeting physically to avoid the COVID-19 outbreak. Online rituals during this pandemic help eliminate the fear and anxiety of religious communities regarding the COVID-19 outbreak, which has claimed many survivors.

1.4.3 Use of Digital-Based Communication Through Social Networks

In the earlier days of COVID-19 in March 2020, Indonesian society, including communities in all research locations, really needed information about COVID-19. How to handle it, but information from official sources was minimal. Some community members, especially those in urban areas who already have access to digital communications, sought information from social media, especially via the WhatsApp site (WA). They received information that was different from one another and mazy, causing confusion and worry. Information that was not necessarily true, many of which were even hoaxes without being filtered first, was spread through WA groups of family members and other communities.

Much of the information about COVID-19 shared in WA groups caused excessive concern and fear among most community members regarding the impact of COVID-19. The conditions at the Malang Regency research location caused the culture of helping each other to suddenly stop. The neighbours around where the COVID-19 patient (Mrs. K) lives were very worried and afraid, they even avoided the road in front of the patient's house. They did not assist Mrs. K and her family who were self-isolating. Therefore, assistance for the patient's needs, especially food and other basic needs, had to be provided by a distant neighbour (Mrs. N) and work colleagues who lived far from the patient's house. The patient's neighbours were also suspicious and worried about Mrs N (50 years), as stated below:

> People don't want to be close to me. In the WA group, I was said to be a COVID spreader because I helped Mrs K [COVID patient] and went to her house. ... I was also kicked out when I went to Mr. A's house, Mrs. K's neighbour... he said ... because I contacted Mrs. K... this made a big splash on WA... Another story when Mrs. K's husband threw rubbish in front of his house... it was also written about on the WA group and reported to the COVID Task Force here.

In mid-to-late 2020, when the knowledge of most of the community increased and more elderly people and community members were detected with COVID-19, community awareness about the impact of COVID-19 and how to suppress its spread also increased. The cultural habit of helping each other with close neighbours in the area was restored.

In mid-2023, communities at all research locations had been connected to virtual space, and most of them carried out digital communication through their social networks. They interacted indirectly with family and other community members using digital platforms. This communication uses digital tools, especially cell phones/smartphones, accompanied by certain applications or social media sites. There were quite a lot of WA community groups, including WA groups of family, neighbours, relatives, or friends who have the same profession/hobby/school background/social and religious activities.

WA was very useful for the community members and other groups to stay in touch, greet each other, and ask how each other was doing before the pandemic. Even though conversations via WA chat were limited, this communication could at least help the community members, especially the elderly, overcome feelings of homesickness and a sense of isolation due to the social restrictions they had to comply with.

> During COVID, my children and grandchildren rarely came to my house, but we still communicated by telephone or chatting via WA... but I still miss them... My grandchildren then taught me how to use video calls...I was very happy... even though I couldn't meet them face to face, I could still see my children and grandchildren on my smartphone... (Mrs. M, 67 years)

Mrs. R (46 years), a COVID-19 survivor said:

> When I got COVID... I isolated myself in the room on the second/top floor of my house... I was very lonely... I couldn't do the activities I usually do.... that's why I was happy when I received chats from my children, neighbours and friends. They tried to calm and comfort me and provide spiritual support, such as getting well soon and keep your spirits up.

WhatsApp is also an important medium for conveying announcements or messages with special material, such as about COVID-19 and its handling, religious/worship activities, and other social activities. For example, announcements from government authorities or religious and traditional leaders regarding regulations for the implementation of worship in places of worship (Mosques, Churches, and Temples/*Pura Banjar*) were distributed via WA RT/RW, WA Muslim and Christian community groups, and WA *Banjar* traditional community groups. Announcements regarding vaccinations and vaccine schedules were distributed via WA RT/RW. Announcements were also distributed through other WA-WA groups, such as announcements of *posyandu* schedules and elderly sports activities through the WA group for the elderly and WA for the *Dasawisma* group.

Community digital communication networks also contribute greatly as a medium for sharing information for the elderly and other community members. The information material is varied, especially information about COVID-19 and its treatment assistance to the elderly or residents who have been exposed to COVID-19 and are undergoing self-isolation. Information includes news about food assistance, coordination of patient family representatives with the head of the RT/RW/management or patient health management task force, transportation to and from the hospital, and/or necessary health equipment such as oxygen cylinders and other equipment, especially for patients who hospitalized, as well as information regarding preparedness and mitigation. Efforts to maintain immunity and health conditions.

1.5 Discussions

This research reveals that the community's social capital in all research locations plays an important role for its members in facing COVID-19 and adapting to disruption during and after the pandemic. These communities apply a combination of socio-cultural practices and digital-based communication through their social networks to deal with the pandemic and maintain their health conditions. With this social capital, they carry out collective activities, working together to help each other overcome problems or instability due to pandemic disruption.

1.5.1 The Role of Socio-cultural and Religious Practices

The communities in all research locations utilize local knowledge and wisdom that they have practiced for a long time and are the legacy of their ancestors. Communities respond to the disruption of the COVID-19 pandemic through coping and adaptation mechanisms to maintain their health conditions during and after the pandemic. The community carries out three significant efforts: (1) mutual help to assist community members who are exposed to the coronavirus and need health assistance related to the socio-economic impact of the pandemic, (2) cultural rituals to prevent the spread of the coronavirus, and (3) support through social and religious/worship/spiritual activities.

1.5.1.1 Assistance to COVID-19 Survivors and Community Members in Need

Mutual help and cooperation in everyday life are socio-cultural practices commonly carried out by people in almost all regions of Indonesia, including in all research locations. This cultural heritage contains the norms and values of togetherness, community concern, and behaviour guidelines (Lestari and Sularso 2020; Saleh Al-Omoush et al. 2020). This habit is, therefore, very necessary for communities in conditions of instability during the pandemic. Communities in Malang, Tabanan, and Wonogiri Regencies, as well as Bekasi and Semarang Cities, were helping their community members exposed to COVID-19 and/or undergoing self-isolation at home or designated isolation places. They tried to meet their food needs and other basic needs in two ways. First, the provision of food or ready-to-eat food was carried out in coordination with the RT/RW (Rukun Tetangga/Rukun Warga or Neighborhood Association/Community Association) with funds sourced from monthly contributions from all RT residents. Second, food provision was provided through individuals, such as neighbours, close friends/relatives, and other residents whose economic conditions were relatively good. In addition, the community (especially community leaders, RT/RW/Banjar caretakers) also coordinates with the families of COVID-19 patients/survivors and the health authorities/COVID-19 Task Force at the village/sub-district level to monitor patient's health condition and assist contact family members who live elsewhere.

In emergency conditions, mutual aid activities were also carried out by community members, especially those who had the same disease, through lending and borrowing medicines that they both consumed. Hidayati et al. (2023) stated that this form of mutual assistance occurred in Bekasi City, where a small number of community members in a city's housing complex, especially those who had illnesses and took the same medication, such as heart medication, high blood pressure, and diabetes.

This activity was carried out by members of the elderly community with limited economic conditions. Before the pandemic, they routinely checked their health at the same health centre (*Puskesmas*), so they received the same medicines. During the pandemic, the elderly's access to health services at *Puskesmas* decreased because it focused more on treating confirmed COVID-19 patients. Due to their economic constraints, these elderlies experienced difficulties gaining access to private healthcare facilities. Consequently, they had to borrow the medicine from friends and return it when they received money, for instance, from children or other family members.

Support from the community in these research locations, according to Stuart and Sundeen (1991), is a problem-centred coping mechanism. Communities with social ties assist COVID-19 survivors and community members who need help. The coping efforts made are positive behavioural responses used by the community to solve the problems faced due to the pandemic. The community actively solves problems through effective coping strategies (Sheidow et al. 2014). This finding also supports the research results of Li and Peng (2021), which revealed a relationship between social support, coping strategies, and worry. Social support influences cognitive, behavioural, and emotional coping efforts, including reducing emotional stress and anxiety among Chinese students at the peak of the COVID-19 pandemic from January to February 2020.

1.5.1.2 Efforts to Suppress the Spread of COVID-19

Apart from helping each other, some community members are also trying to prevent and reduce the spread of COVID-19 through cultural rituals and limiting religious and other social activities. This cultural practice is in accordance with the WHO/EHA (1999) concept of non-standard coping strategies because they are based on the culture and needs of each individual or community group. Some communities in Malang Regency, especially communities in Singosari and Pakis Districts, perform the "reject disaster" ritual to "drive away" COVID-19. They carry out this ritual based on the perception that COVID-19 is a "pageblug", namely a disease outbreak like the mass disease that previously hit their areas. Rituals are executed by combining cultural practices and religious activities together. The community that lives nearby puts out a fire in front of the house and offers prayers which are then followed by eating serai (cakes made from rice/

flour) and *ketupat* (food made from rice) together in the afternoon—sunset. This method of dealing with the "pageblug" outbreak is also widely used by communities in East Java. Nugraha (2020) and Rumilah et al. (2020) stated that apart from offering prayers, the community also cleans heirlooms, consumes seven-coloured *lodeh* vegetables, and sprays disinfectant from natural ingredients because they believe that this tradition is able to eliminate the COVID-19 outbreak.

In the context of COVID-19, cleaning the environment using herbal (natural) disinfectants and lighting fireplaces is an effort to suppress the spread of the virus, while praying is a spiritual activity that aims to manage the emotions of the community and ask Allah "the Creator" to protect them from the coronavirus and so that they can living life and livelihood during the pandemic. However, this ritual activity is carried out together with community members, causing crowds, because it is not in accordance with health protocols and has the potential to accelerate the transmission of the virus. The community's trust in this "pageblug" is decreasing with the increasing information they receive, but until the research was conducted in May–June 2022, some elderly people and community members, especially those living in villages/villages in Malang Regency, still lacked confidence in COVID-19 and because of this, they do not comply with the social restrictions and health protocols imposed by the government and the COVID-19 Task Force.

1.5.1.3 Religious and Spiritual Roles

The COVID-19 pandemic has also had a major impact on religious practices (Coppola et al. 2021; Moroń et al. 2021; Moghadam et al. 2021). Society makes efforts through spiritual approaches or increasing religiosity to build resilience during the pandemic (Gavriluță et al. 2022a, b; Pabbajah et al. 2022). They perform religious rituals such as reading the Koran, prayers and prayers to improve prosperity amidst instability (Baidowi et al. 2021).

Worship activities in places of worship (mosques, temples, churches) are important activities that are routinely carried out by most communities in all research locations. Worship activities are considered increasingly important for community members, especially the elderly, during the COVID-19 pandemic as a spiritual effort to reduce feelings of anxiety, worry, and fear of contracting the coronavirus and its impact on their lives and livelihoods. However, social restrictions and health protocols hamper community worship activities, although these efforts are very necessary to maintain their health.

The role of religious and cultural leaders who are respected and obeyed by communities is very important in increasing community awareness to comply with the rules, even though a small number of community members, especially in villages, still secretly or openly practice religious activities as usual because they still do not believe in the existence of COVID-19, believing that God will protect them. Religious figures can provide positive advice and provide encouragement and motivational support for COVID-19 survivors.

1.5.2 The Role of Social Networks in Accelerating the Transformation of the Use of Digital Communications During and Post-pandemic

The COVID-19 pandemic has accelerated and increased the use of digital communication technologies by communities in all research locations, especially in urban areas. In the earlier times of the pandemic, the need to obtain information related to COVID-19 was very high, but information from official sources was very limited and interaction with family, friends, and relatives to gain information offline and carry out socio-economic activities was also limited due to the implementation of social restrictions and health protocols. This condition encourages communities to increase their use of social media so they can access information and communicate to meet their health and socio-economic needs. Most of the community, including the elderly, are "forced" to use cell phones/smartphones and learn how to operate them, especially through family and community WhatsApp groups.

Communities in all research locations intensify and/or expand the use of digital communication through social networks that have been formed with varying goals according to the needs of each individual and group. Digital communication activities carried out by this community are a form of coping and adaptation mechanism for individuals, families, and communities to assist them in responding to problems and adapting to pandemic disruption. According to Finlay et al. (2021), this mechanism is a cognitive and behavioural strategy, and the efforts they make involve feelings, thoughts, and actions (Lazarus and Folkman 1984).

The coping strategies of community members through digital communication networks in all research locations, according to Stuart and Sundeen (1991), are problem-centred, namely efforts to overcome problems and improve the situation by asking for help or collaborating with family, friends, and community groups. WhatsApp community groups are an important medium for conveying information about the conditions being faced, such as about community members who have been exposed to COVID-19 and the health measures that are being taken, restrictions on religious/worship activities, and other social activities.

The community digital communication network also makes a big contribution as a medium for sharing information about food assistance for residents who have been exposed to COVID-19 and are undergoing self-isolation, and mitigation efforts to maintain immunity and health conditions. Apart from that, the community also provides emotional support. WA is very useful for group members to communicate, greet each other, and ask how each other is doing. Even though conversations via WA chat are limited, this communication can at least help community members overcome feelings of longing and isolation during the pandemic. Community members also send spiritual or psychological messages that provide support and encouragement to the elderly or community members who have been exposed to COVID-19, such as prayers for healing, advice/encouragement to drink milk and eat nutritious food, do exercise to get better results. This spiritual or psychological support is very helpful for patients who generally experience boredom during self-isolation or stress due to the illness.

The results of this research support several research results which reveal community efforts, including establishing social ties, intellectual formation, and spiritual support or increasing religiosity (Gavriluță et al. 2022a, b; Pabbajah et al. 2022). First, establishing social bonds, as shown by Ramkissoon (2022) through his research, found that social bonding has helped foster a more positive attitude and made someone more resilient. So that both residents who are affected and not affected are equally prosperous and healthier mentally, physically, and emotionally. Second, through intellectual formation, for example, by reading true news about the pandemic and not being influenced by fake news circulating widely. And third, with a spiritual approach or increasing religiosity. Religious rituals such as reading the Koran, prayers, prayers, and hizib from a psychological perspective are part of mysticism, symbolization, and religious rhetoric to improve welfare in the midst of instability (Baidowi et al. 2021). The role of religious and spiritual figures is important in providing advice and motivational support to the community, especially COVID-19 survivors (Hatabu et al. 2021).

1.6 Conclusion

This article reveals qualitative evidence that contributes to knowledge about the subjective reality of the important roles of the community social capital in supporting community members to overcome uncertainty and stressors during the COVID-19 pandemic. Various efforts have been carried out by the community to build resilience in dealing with and adapting to the pandemic. They utilize the social capital that has developed in their region as a coping and adaptation strategy. The community applies coping mechanisms to protect and care for the community members based on two main strategies, namely local knowledge and wisdom and community virtual communication. The community practices the social support system implemented in the area to assist its members in surviving the COVID-19 crisis, as well as maintain their health condition and daily socio-economic activities during and after the pandemic.

Communities with their social capital work together to assist their members, ensure the availability of food during self-isolation, distribute aid for basic needs, coordinate with health authorities/local COVID-19 Task Forces, especially in urban areas, and maintain community activities in accordance with health protocols. These socio-cultural practices need to be maintained to increase preparedness and reduce the risk of community members in anticipating future pandemics and other relevant diseases.

The role of community social networks through digital communication is also effective in assisting communities deal with and adapt to the pandemic, especially in urban areas. This transformation in the use of virtual communication needs to be fostered with guidance and increasing digital literacy among individuals, families, and communities. This effort is very important because some community members, particularly elderly people in rural areas, are still not familiar with the internet and digital communication technology, and some other elderly people do not have cell phones/smartphones. Increasing digital literacy is also needed so that they can sort out information circulating in virtual space by checking and verifying the veracity of information from trustworthy sources and not spreading information that is not necessarily true or is a hoax.

Community support creates a social infrastructure that has the potential to have practical implications for improving the quality of life of elderly people who are resilient in facing the shock of the COVID-19 pandemic and recover with better health conditions and social life. Therefore, social infrastructure needs to be strengthened with policies that support social cohesion and relevant socio-cultural practices in dealing with the pandemic through increasing capacity and providing community access to health services and facilities, as well as socio-cultural and religious practices that could adapt to the new normal conditions and anticipate future pandemics.

References

Abdullah I (2020) COVID-19: threat and fear in Indonesia. Psychol Trauma Theory Res Pract Policy 12(5):488–490. https://doi.org/10.1037/tra0000878

Aguilar-Latorre A, Oliván-Blázquez B, Porroche-Escudero A, Méndez-López F, García-Gallego V, Benedé-Azagra B, Magallón-Botaya R (2022) The impact of the COVID-19 lockdown on depression sufferers: a qualitative study from the province of Zaragoza, Spain. BMC Public Health 22(1):1–13. https://doi.org/10.1186/s12889-022-13083-2

Al Mubarroh NR, Susanto IH, Mustar YS (2021) Physical activity and aspects of concern for the elderly during the Covid-19 pandemic. Altius Jurnal Ilmu Olahraga Dan Kesehatan 10(1):97–111. https://doi.org/10.36706/altius.v10i1.14140

Al-Omousha KS, Simón-Moyab V, Sendra-García J (2020) The impact of social capital and collaborative knowledge creation on e-business proactiveness and organizational agility in responding to the COVID-19 crisis. J Innov Knowl 5(4):100260. https://doi.org/10.1016/j.jik.2020.10.002

Alshurafat H, Al Shbail MO, Masadeh WM, Dahmash F, Al-Msiedeen JM (2021) Factors affecting online accounting education during the COVID-19 pandemic: an integrated perspective of social capital theory, the theory of reasoned action and the technology acceptance model. Educ Inf Technol 26(6):6995–7013. https://doi.org/10.1007/s10639-021-10550-y

Baidowi A, Salehudin A, Mustaqim A, Qudsy SZ, Hak N (2021) Theology of health of quranic pesantren in the time of COVID-19. HTS Teologiese Stud Theol Stud. https://doi.org/10.4102/hts.v77i4.6452

Bhardwaj R (2021) Risk, causation and containment of Covid-19 pandemic in India: a sociological interpretation. Int Rev Sociol 32(1):10–28. https://doi.org/10.1080/03906701.2021.1996758

Bidzan-Bluma I, Bidzan M, Jurek P, Bidzan L, Knietzsch J, Stueck M, Bidzan M (2020) A polish and German population study of quality of life, well-being, and life satisfaction in older adults during the COVID-19 pandemic. Front Psych 11(585813):1–9. https://doi.org/10.3389/fpsyt.2020.585813

Borgonovi F, Andrieu E (2020) Bowling together by bowling alone: social capital and COVID-19. Soc Sci Med. https://doi.org/10.1016/j.socscimed.2020.113501

Borgonovi F, Andrieu E, Subramanian SV (2021) The evolution of the association between community level social capital and COVID-19 deaths and hospitalizations in the United States. Soc Sci Med 278(113948):1–13. https://doi.org/10.1016/j.socscimed.2021.113948

Borkowska M, Laurence J (2021) Coming together or coming apart? Changes in social cohesion during the Covid-19 pandemic in England. Eur Soc 23(S1):S618–S636. https://doi.org/10.1080/14616696.2020.1833067

Brown EM, Fernald LCH, Hamad R, Hoskote M, Jackson KE, Gosliner W (2022) Pandemic-related socio-economic disruptions and adverse health outcomes: a cross-sectional study of female caregivers. BMC Public Health 22(1):1–11. https://doi.org/10.1186/s12889-022-14287-2

Caliskan A, Ezilmez B (2021) A theoretical review on social capital. Hitit J Econ Polit 1(1):18–30

Carver CS, Scheier MF, Weintraub JK (1989) Assessing coping strategies: a theoretically based approach. J Person Soc Psychol 56(2):267–283. https://doi.org/10.1037/0022-3514.56.2.267

Chigwedere OC, Sadath A, Kabir Z, Arensman E (2021) The impact of epidemics and pandemics on the mental health of healthcare workers: a systematic review. Int J Environ Res Public Health 18(13):6695. https://doi.org/10.3390/ijerph18136695

Coppola I, Rania N, Parisi R, Lagomarsino F (2021) Spiritual well-being and mental health during the COVID-19 pandemic in Italy. Front Psych 12(626944):1–15. https://doi.org/10.3389/fpsyt.2021.626944

Cowan HR (2020) Is schizophrenia research relevant during the COVID-19 pandemic? Schizophr Res 220:271–272. https://doi.org/10.1016/j.schres.2020.04.002

Doloksaribu M, Verayanti Manalu N (2021) Level of knowledge and anxiety of the elderly about the covid-19 pandemic disease in Uluan district, Toba Regency. Jurnal Sosial Sains 1(12):1631–1636. https://doi.org/10.36418/sosains.v1i12.280

Elgar FJ, Stefaniak A, Wohl MJA (2020) Since January 2020 Elsevier has created a COVID-19 resource centre with free information in English and Mandarin on the novel coronavirus COVID-19. The COVID-19 resource centre is hosted on Elsevier Connect, the company's public news and information 263(113365). https://doi.org/10.1016/j.socscimed.2020.113365

Ellison NB, Steinfield C, Lampe C (2007) The benefits of facebook "friends:" Social capital and college students' use of online social network sites. J Comput Mediat Commun 12(4):1143–1168. https://doi.org/10.1111/j.1083-6101.2007.00367.x

Evans D, Borriello GA, Field AP (2018) A review of the academic and psychological impact of the transition to secondary education. Front Psychol 9(August 29):1482. https://doi.org/10.3389/fpsyg.2018.01482

Finlay JM, Kler JS, O'Shea BQ, Eastman MR, Vinson YR, Kobayashi LC (2021) Coping during the COVID-19 pandemic: a qualitative study of older adults across the United States. Front Public Health 9(April):1–12. https://doi.org/10.3389/fpubh.2021.643807

Ford M, Ward K (2021) COVID-19 in Southeast Asia: implications for workers and unions. J Ind Relat 63(3):432–450. https://doi.org/10.1177/00221856211000097

Gavriluță C, Dalban CM, Ioan BG (2022a) Educational, emotional, and social impact of the emergency state of COVID-19 on Romanian university students. Int J Environ Res Public Health 19(7):3990. https://doi.org/10.3390/ijerph19073990

Gavriluță C, Dalban CM, Ioan BG (2022b) Educational, emotional, and social impact of the emergency state of COVID-19 on Romanian university students. Int J Environ Res Public Health. https://doi.org/10.3390/ijerph19073990

Haber A, Runyon RP (1984) Psychology of adjustment. The Dorsey Press, New York

Hatabu A, Mao X, Zhou Y, Kawashita N, Wen Z, Ueda M, Takagi T, Tian YS (2021) Knowledge, attitudes, and practices toward COVID-19 among university students in Japan and associated factors: an online cross-sectional survey. PLoS ONE 15(December 12):1–19. https://doi.org/10.1371/journal.pone.0244350

Hidayati D (2020) Profile of residents who are confirmed positive for COVID-19 and died: cases of Indonesia and DKI Jakart. J Indones Popul Spec Ed Demographics COVID-19, July:93–100

Hidayati D, Widayatun, Situmorang A (2023) The importance of family and community approach in reducing the risk faced by the elderly during the COVID-19 pandemic. J Dis Res. https://doi.org/10.20965/jdr.2023.p0415

Horwitz IM, Lascar S (2021) Ties in tough times: how social capital helps lower-income Jewish parents weather the economic hardship of COVID-19. Contemp Jew 41(1):161–183. https://doi.org/10.1007/s12397-021-09385-z

Ilukor J, Joyce A, Okiror SP (2022) The impacts of COVID-19 on cattle traders and their response in agro-pastoral and pastoral regions in Uganda: a case of Karamoja and Teso cattle traders. Pastorialisme 12(1):18

Kusumawati D, Saiffudin (2020) Pengaruh Persepsi Harga dan Kepercayaan Terhadap Minat Beli Secara Online Saat Pandemi Covid-19 Pada Masyarakat Millenia di Jawa Tengah 21(1):1–9

Lazarus RS, Folkman S (1984) Stress, appraisal, and coping. Springer, New York

Lestari P, Sularso S (2020) The COVID-19 impact crisis communication model using Gending Jawa local wisdom. Int J Commun Soc 2(1):47–57. https://doi.org/10.31763/ijcs.v2i1.150

Li S, Xu Q (2022) Family support as a protective factor for attitudes toward social distancing and in preserving positive mental health during the COVID-19 pandemic. J Health Psychol 27(4):858–867. https://doi.org/10.1177/1359105320971697

Lin Y, Xiao H, Lan X, Wen S, Bao S (2020) Living arrangements and life satisfaction: mediation by social support and meaning in life. BMC Geriatr 20(1):1–8. https://doi.org/10.1186/s12877-020-01541-8

Liu K, Chen Y, Lin R, Han K (2020) Clinical features of COVID-19 in elderly patients: a comparison with young and middle-aged patients. J Infect 80(6):e14–e18. https://doi.org/10.1016/j.jinf.2020.03.005

Loades ME, Chatburn E, Higson-Sweeney N, Reynolds S, Shafran R, Brigden A, Linney C, McManus MN, Borwick C, Crawley E, Barbieri V, Wiedermann CJ, Kaman A, Erhart M, Piccoliori G, Plagg B, Mahlknecht A, Ausserhofer D, Engl A, Ravens-Sieberer U (2022) Rapid systematic review: the impact of social isolation adolescents in the context of COVID-19. Int J Environ Res Public Health 19(11):1218–1239

Marchi M, Magarini FM, Chiarenza A, Galeazzi GM, Paloma V, Garrido R, Ioannidi E, Vassilikou K, De Matos MG, Gaspar T, Guedes FB, Primdahl NL, Skovdal M, Murphy R, Durbeej N (2022) Experience of discrimination during COVID_19 pandemic: the impact of public health measures and psychological distress among refugees and other migrants in Europe. BMC Public Health 22(942):1–14. https://doi.org/10.1186/s12889-022-13370-y

Marliana T, Keliat BA, Kurniawan K (2022) Mental health and psychosocial support online-based services to improve elderly integrity and reduce loneliness during the pandemic covid-19. Malays J Med Health Sci 18:1–6

Michelen M, Jones N, Stavropoulou C (2020) In patients of COVID19, what are the symptoms and clinical features of mild and moderate cases? The Centre for Evidence-Based Medicine. https://www.Cebm.Net/COVID-19/

Moghadam MT, Taati B, Paydar Ardakani SM, Suzuki K (2021) Ramadan fasting during the COVID-19 pandemic; observance of health, nutrition and exercise criteria for improving the immune system. Front Nutr 7(570235):1–10. https://doi.org/10.3389/fnut.2020.570235

Moroń M, Biolik-Moroń M, Matuszewski K (2021) Alterations in religious rituals due to COVID-19 could be related to intragroup negativity: a case of changes in receiving holy communion in the roman catholic community in Poland. Religions 12(240):1–22. https://doi.org/10.3390/rel12040240

Ngadi N, Meliana R, Purba YA (2020) Dampak Pandemi Covid-19 Terhadap Phk Dan Pendapatan Pekerja Di Indonesia. Jurnal Kependudukan Indonesia 2902:43. https://doi.org/10.14203/jki.v0i0.576

Nugraha AS (2020) Sociology of local wisdom in facing the covid-19 pandemic: a literature review. Jurnal Sosietas 10(1):745–753

Okuyama J, Seto S, Fukuda Y, Funakoshi S, Amae S, Onobe J, Izumi S, Ito K, Imamura F (2021) Mental health and physical activity among children and adolescents during the COVID-19 pandemic. Tohoku J Exp Med 253(3):203–215. https://doi.org/10.1620/tjem.253.203

Pabbajah M, Muhammad Said N, Taufiq Hidayat Pabbajah M, Jubba H (2022) Deauthorization of the religious leader role in countering Covid-19: perceptions and responses of Muslim societies on the Ulama's policies in Indonesia. Iournal of Religius and Cross Culturual Studies 6(3):341–352. https://doi.org/10.15575/rjsalb.v6i3.16789

Park S, Cho J (2022) Young people are medically invulnerable to COVID-19 but vulnerable in the labor market: Korean evidence. Heal Econ Rev 12(1):1–16. https://doi.org/10.1186/s13561-022-00360-4

Pascaru M (2021) Community and social control during the covid-19 pandemic. J Commun Posit Pract 3(21):3–16. https://doi.org/10.35782/JCPP.2021.3.01

Pitas N, Ehmer C (2020) Social capital in the response to COVID-19. Am J Health Promot 34(8):942–944. https://doi.org/10.1177/0890117120924531

Prayitno G, Hayat A, Efendi A, Auliah A, Dinanti D (2022) Structural model of community social capital for enhancing rural communities adaptation against the COVID-19 pandemic: empirical evidence from Pujon Kidul Tourism Village, Malang Regency, Indonesia. Sustainability (Switzerland). https://doi.org/10.3390/su141912949

Promislow DEL, Phil D (2020) A geroscience perspective on COVID-19 mortality. J Gerontol A Biol Sci Med Sci 75(9):e30–e33

Rahmawati YSFFA, Hati EM, Roziqin A (2021) Local wisdom in facing Covid-19: study of resilient kampong in East Java. J Resilient Econ 47(1):1–12

Ramkissoon H (2022) COVID-19 adaptive interventions: implications for wellbeing and quality-of-life. Front Psychol 17(13):810951. https://doi.org/10.3389/fpsyg.2022.810951

Razai MS, Oakeshott P, Kankam H, Galea S, Stokes-lampard H (2020) Mitigating the psychological effects of social isolation during the covid-19 pandemic. BMJ Open 21(369):1–5. https://doi.org/10.1136/bmj.m1904

Reger M, Stanley I, Joiner T (2020) Suicide mortality and coronavirus disease 2019—A perfect storm? JAMA Psychiat 77(11):1093–1094. https://doi.org/10.1001/jamapsychiatry.2020.1060

Salman A, Al-Ghadban F, Sigodo KO, Taher AK, Chun S (2021) The psychological and social impacts of curfew during the covid-19 outbreak in Kuwait: a cross-sectional study. Sustainability (Switzerland) 13(15):1–16. https://doi.org/10.3390/su13158464

Sarafino E, Smith T, DeLongis A, King D (2015) Health psychology: biopsychosocial interactions. Wiley, London

Sari MT, Daryanto D (2021) Elderly's health and happy education (smart) during the covid-19 pandemic at Tempino Public Health Center, Muaro Jambi. LOGISTA Jurnal Ilmiah Pengabdian Kepada Masyarakat 5(1):35–41. https://doi.org/10.25077/logista.5.1.35-41.2021

Shahid Z, Kalayananitra R, McClafferty B, Kepko D, Ramgobin D, Patel R, Anggarwal S, Vunnam R, Sahu N, Bhatt D, Jones K, Pharm D, Golamari R, Jain R (2020) COVID-19 and older adults: what we know. J Am Geriat Soc 68(5):926–929. https://doi.org/10.1111/jgs.16472

Shanahan L, Steinhoff A, Bechtiger L, Murray AL, Nivette A, Hepp U, Ribeaud D, Eisner M (2022)

Emotional distress in young adults during the COVID-19 pandemic: evidence of risk and resilience from a longitudinal cohort study. Psychol Med 52(5):824–833. https://doi.org/10.1017/S003329172000241X

Sheidow AJ, Henry DB, Tolan PH, Strachan MK (2014) The role of stress exposure and family functioning in internalizing outcomes of urban families. J Child Fam Stud 23(8):1351–1365. https://doi.org/10.1007/s10826-013-9793-3

Stansfeld S, Candy B (2006) Psychosocial work environment and mental health—a meta-analytic review. Scand J Work Environ Health 32(6):443–462. https://doi.org/10.5271/sjweh.1050

Sundeen, Stuart (1991) Pocket guide to psyhiatric nursing, 3rd edn. The Mosby Company Toronto, London

Van Bavel JJ, Baicker K, Boggio PS, Capraro V, Cichocka A, Cikara M, Crockett MJ, Crum AJ, Douglas KM, Druckman JN, Drury J, Dube O, Ellemers N, Finkel EJ, Fowler JH, Gelfand M, Han S, Haslam SA, Jetten J, Weeden KA (2020) COVID-19 pandemic response. Nat Hum Behav 4(5):460–471. https://doi.org/10.1038/s41562-020-0884-z

Wang H, Paulson KR, Pease SA, Watson S, Comfort H, Zheng P, Aravkin AY, Bisignano C, Barber RM, Alam T, Fuller JE, May EA, Jones DP, Frisch ME, Abbafati C, Adolph C, Allorant A, Amlag JO, Bang-Jensen B, Murray CJL (2022) Estimating excess mortality due to the COVID-19 pandemic: a systematic analysis of COVID-19-related mortality, 2020–21. Lancet 399(10334):1513–1536. https://doi.org/10.1016/S0140-6736(21)02796-3

World Health Organization (WHO) (2008) Report from the commission on social determinants of health

World Health Organization (2020) Medical certification, ICD mortality coding, and reporting mortality associated with COVID-19, vol 19 (June)

Xiong J, Lipsitz O, Nasri F, Lui LMW, Gill H, Phan L, Chen-Li D, Iacobucci M, Ho R, Majeed A, McIntyre RS (2020) Impact of COVID-19 pandemic on mental health in the general population: a systematic review. J Affect Disord 1(277):55–64. https://doi.org/10.1016/j.jad.2020.08.001

Zhang D, Lu S, Morse S, Liu L (2021) The impact of COVID-19 on business perspectives of sustainable development and corporate social responsibility in China. Environ Dev Sustain 24(6):8521–8544. https://doi.org/10.1007/s10668-021-01798-y

2 Towards the Development of an Ideal Risk Communication by Acknowledging Local Culture

Mochammad Wahyu Ghani

Abstract

Risk communication is a key element of disaster management. Communities living in disaster-prone areas often rely on myths, metaphors, and various spiritual beliefs to communicate risks, indicating an intergenerational transfer of risk knowledge and communication. An effective risk communication system in such areas begins with a well-structured collective memory and the transfer of knowledge. However, disaster early warning efforts are often carried out using a top-down approach that does not acknowledge the risk culture already existing within the community. This approach fails to effectively convey disaster risk because people already have their own perceptions. While national governments should contribute essential elements to disaster risk management, local risk communicators must remain flexible in delivering information and recognize the diverse worldviews of each community. Governments cannot overlook the local knowledge, attitudes, and motivations present in communities while delivering disaster risk communication. Instead, a multi-way dialogue should be established, allowing governments and local communities to learn from each other and improve overall disaster risk management.

Keywords

Risk communication · Local culture · Risk management

2.1 Introduction

Indonesia is geographically one of the countries with the highest disaster potential in the world. It is home to 127 active volcanoes, accounting for approximately 13% of the world's active volcanoes, making it the country with the largest number of volcanoes globally (Cottrell 2015; BNPB 2016). Additionally, Indonesia is located at the convergence of the Indo-Australian, Eurasian, Pacific, and Philippine tectonic plates, resulting in a very high risk of disasters such as earthquakes, volcanic eruptions, and tsunamis (Fuady et al. 2021; Pribadi et al. 2021). This geographical setting has contributed to Indonesia's long history of major disasters that have had global impacts, including influencing world climate change. Notable examples include the eruptions of Mount Krakatau and Mount Tambora in the 1800s (Rampino and Self 1982; Gleckler et al. 2006).

M. W. Ghani (✉)
Research Center for Population - The National Research and Innovation Agency (BRIN), Jakarta, Indonesia
e-mail: moch039@brin.go.id

D. Hidayati et al. (eds.), *Examining Disaster Risk Reduction in Indonesia*,
Sustainable Development Goals Series, https://doi.org/10.1007/978-981-96-4182-6_2

The country's vulnerability to disasters has long been recognized by its people. Consequently, knowledge about disasters has been transmitted across generations through folklore, myths, and philosophical teachings. In West Sumatra, a region with a history of significant earthquakes and volcanic eruptions, the severity of these events is reflected in the folklore of Bujang Sambilan, which is closely associated with seismic and volcanic activities (Tanhas et al. 2017; Anwar 2020). Similarly, Bondowoso Regency preserves a myth about the struggle between Dhamar Wulan and Menak Jinggo, wherein the lava of Mount Raung is believed to represent their flowing blood (Purnomo and Kurniawan 2019).

The city of Palu, known for its earthquake- and tsunami-prone topography, transmits its local understanding of these hazards through its name. Derived from the word "Topalu'e," meaning "uplifted land," the city's name itself serves as a reminder of the geophysical forces shaping the region (Rahman et al. (Eds.), 2021, pp. 25–37; Allokendek and Ellisa 2022). In addition, village elders and community leaders throughout the country play a crucial role in passing down this local wisdom, thereby enriching disaster education and preparedness efforts (Septiana et al. 2019).

To survive in nature, the ancestors of every human culture used myths, metaphors, and rituals to serve as reminders of the potential risks they would face (Heath and O'Hair 2020). In such a context, communication becomes essential to transferring knowledge about potentially dangerous situations that can be passed down to each generation below them (Lin et al. 2020). The scientific community then seeks to translate risk communication knowledge as part of risk management and perception that grows to "improve the quality of risk decisions through better communication" (Palenchar and Heath 2002, p. 129; Sellnow et al. 2009, p. 5).

The government's risk communication strategy is crucial for conveying vital risk information to communities before, during, and after a disaster as part of preventive public education efforts (Hilburn and Zawisza 2019). To achieve this objective, the government should collaborate with local communities to share knowledge about the importance of disaster mitigation within the area. Such collaborative engagement is essential for identifying local community knowledge that can inform policy formulation and for disseminating this knowledge to other relevant regions (Mercer et al. 2010). Furthermore, collaborative risk communication is critical to bridging disaster risk reduction policies among the government, academia, and local communities, ensuring a cohesive and inclusive approach.

Indigenous Indonesians possess distinct knowledge systems and communication styles that differ significantly from Western approaches. Scientists and practitioners often regard local knowledge as inferior to scientific knowledge (Mercer et al. 2010). Disaster science, which is typically dominated by Western paradigms, prioritizes explicit, logical, and evidence-based information. In contrast, communication in Indonesia is more effective when employing high-context methods, where messages are deeply embedded and appear "implicit" (Nishimura et al. 2008; Gupta and Sukamto 2020; Juneza et al. 2023). This preference for a tongue-in-cheek, life-philosophical style of communication is deeply ingrained in Indonesian culture. Consequently, disaster knowledge is often preserved through myths and fairy tales, which serve as culturally resonant vehicles for intergenerational knowledge transfer (Purnomo and Kurniawan 2019; Prawoto and Octavia 2021; Kurnio et al. 2021).

In practice, the government's approach to disaster preparedness and risk communication is typically top-down, relying on Western scientific knowledge without adequately considering local communities' knowledge. The use of "scientifically proven" expertise often underpins the government's communication with communities. As a result, it is unsurprising that people are sometimes reluctant to evacuate during disasters such as volcanic eruptions and landslides, even when warnings are issued (Schwartz-Marin et al. 2022; Burrows et al. 2021). The effective application of scientific knowledge within local communities cannot succeed through conventional top-down processes alone. Instead, it requires the collaborative construction of shared knowledge that integrates both scientific and local insights (da Cunha et al. 2020).

Budget constraints and bureaucratic complexities hamper egalitarian risk communication programmes with local communities. The government still needs risk communication to exchange local knowledge with scientific facts. More flexible research bureaucracy and low literacy of local knowledge are needed for the government, which still considers the management of research organizations to be the same as the management of ordinary official travel (Burhani et al. 2021; Syahputra 2019). Conducting educational or empowerment meetings with communities requires substantial financial resources due to practical needs such as paying discussion participants and providing refreshments. Not to mention the budget orientation of disaster preparedness and mitigation programmes, which are always ineffective due to lack of funding allocation (Tedjokusumo 2023). Not surprisingly, the gap between local knowledge and scientific facts in disseminating risk communication in Indonesia is wide open.

Local stories such as fairy tales, myths, and philosophies of life are essential to understand. They remain in people's memories and can be narrated as part of life to anticipate future disasters (Prawoto and Octavia 2021; Nagamatsu et al. 2021). Should we use local knowledge as a basis for the risk communication messages we want to spread? Are myths and folklore still valuable in this modern age? This study examines a range of myths and local knowledge from several sites in Indonesia, specifically Solok District and Palu City, as the site of informant interviews and observations. The study seeks to understand the use of local knowledge and disaster culture to effectively communicate disaster risk.

2.2 Methods

2.2.1 Research Design

This study employs a literature review combined with a qualitative phenomenological research paradigm. A literature review provides an overview of a specific research question or problem, serving to assess, summarize, and synthesize the issues under investigation (Gerring 2016; Vagle 2018, p. 127; Snyder 2019). The phenomenological paradigm is particularly relevant for studying local knowledge passed down through generations. Without engaging directly with the phenomenon, a literature review alone risks "sedimenting" an ongoing phenomenon, reducing its dynamic nature to static knowledge (Vagle 2018, p. 127). Phenomenology, as a humanistic tradition, emphasizes human experience and our capacity to connect with the emotions and perspectives of others (Bernard 2006, p. 23; McIntosh and Wright 2019). This focus on human experience is especially critical in disaster phenomenology, which explores myths as narratives of origins, whether pertaining to early events or social phenomena (Basher and Ono 2022). In this case, myths are not dismissed as untruths but are instead viewed as social phenomena—assumptions and possibilities shaped by historical beliefs and ideologies (Aguirre 2020).

2.2.2 Data Collection

The research was conducted from 2021 to 2023. The literature review focused on finding similar data on disaster risk communication in the Indonesian region. The literature review included five reputable scientific articles focusing on disaster risk communication conducted by local communities and five articles on disaster risk communication conducted by the government. The semi-systematic review study was chosen to identify research topics and understand the picture of disaster risk communication research traditions not tied to a specific scientific field (Snyder 2019; Laurilia-Pant et al. 2023). In the qualitative approach, in addition to the literature review, data collection included interviews and researcher observation.

Observations were conducted in Solok Regency as one of the first steps of research activities, which is an activity of daily observation to be used as a source of research data (Bordens and Abbott 2018, p. 59). Researchers often visit West Sumatra, especially Solok Regency, because their families have lived around Lake Singkarak since ancestral times. According to Denzin and Lincoln (2018, p. 922), ethics and closeness must be prioritized in observation activities by validating that researchers are part of the community.

Interviews were conducted in a semi-structured design and unstructured interviews. The semi-structured interviews were divided into three main themes regarding disseminating disaster risk communication knowledge: (1) preparatory communication before a disaster, (2) communication during a disaster, and (3) communication after a disaster occurs. The researcher prepared the semi-structured interview as a guideline and loosely tailored it according to the need for informants' readiness to answer questions (Trainer and Graue 2018; Shafie et al. 2022).

Semi-structured interviews were conducted by researchers in interviewing informants from Palu City and Solok Regency via the Internet. Interviews online offer informant flexibility and convenience and reduce travel costs to explore data whenever researchers can (Jankowski and Selm 2005; Archibald et al. 2019; Lobe et al. 2020). In addition, unstructured interviews were also conducted by researchers when making observations in Solok Regency. In unstructured interviews, the researcher has a general idea of the issues to be discussed but has yet to have a predetermined sequence of questions (Bordens and Abbott 2018, p. 280).

2.3 Literature Review

Conventional definitions of risk communication always focus on the information source's intentionality and the information's quality (Plough and Krimsky 1987, pp. 4–10). Risk communication messages are usually linear and produced by the government or other credible organizations. For example, during the COVID-19 outbreak, the WHO advised each country's government to pay attention to communicators (government representatives) to inform the public effectively. The government's ability to communicate effectively with its people is critical. Failure to communicate well leads to loss of confidence in government, damaging economic consequences and, in the worst cases, loss of life (WHO 2020).

For this reason, *risk communication* is always defined as "the exchange of scientific information" aimed at building bridges between interested parties, in this case, the government and society, regarding "health or environmental risks" (Kar and Cohran 2019, pp. 65–66). In other words, risk communication delivered through linear government channels will only succeed with an ongoing process of multidirectional dialogue about hazard events and disaster risk among affected communities (Grabill and Simmons 1998; Marchezini et al. 2017). In this process, risk communication carried out by the government is only sometimes effective because the government is mainly preoccupied with its agenda and believes that dialogue sometimes does not produce anything (Kinoshita 2014). Government risk communication is usually based on linear channels and the absence of an intensive dialogue process of pre-disaster preparedness (Cole and Fellows 2008).

Research on disasters has shown that effective risk communication needs to take into account (1) the community's perceived risk, its emergency response, and how much of the community's population is affected; (2) the social, economic, cultural, and political factors of the affected population; (3) the previous experiences of the affected population; and (4) the tendency of the affected population to seek information from other sources (Cole and Fellows 2008; Marchezini et al. 2017; Kar and Cohran 2019, pp. 65–66; Coombs 2020).

From the existing concept of risk communication, socio-cultural factors and local knowledge have a role to play in making risk communication messages delivered by specific stakeholders more effective in the community. Therefore, the government must recognize the local knowledge in the community before providing official disaster risk communication to the community (Kar and Cohran 2019, pp. 211–213; Tedjokusumo 2023). There should be a multidirectional dialogue channel to ensure that the government can provide the best information after knowing the information or local knowledge that exists in the community. Certain cultural groups may have different perceptions of what constitutes a risk or crisis, and they may also respond differently to messages about impending or ongoing risks and crises (Sellnow et al. 2009, p. 38) (Fig. 2.1).

Risk communicators' efforts to get the public to comply with risk warnings are based on the ethnocentric field, which influences their perceptions of what is or is not relevant to their circumstances (Littlefield and Cowden 2006; Sellnow et al. 2009, p. 38). The ethnocentric environment can be described as concentric circles of family, community, region, state (could be province), nation and world, radiating from the individual as the centre of the environment (Sellnow et al. 2009, p. 38). The closer to the circle's centre, the more focused the individual is on the risk or crisis message prepared by the communicator. The further away from the circle's centre, the less interest the individual has in the message and the less direct control the communicator has in capturing the public's attention.

Stakeholders or disaster risk communicators need to be aware that each society is different and that our society's norms are what is known as a "worldview" or can be interpreted literally as a view of the world that is not necessarily the same as others (Groot et al. 2020; Palmlund 2020). Some communication scholars suggest that value orientations and norms contribute to the "worldview" of a particular society's culture and have a strong influence on how members of that culture perceive and respond to communication (Sellnow et al. 2009, p. 44). In addition, because recipients of risk messages have different perceptions, beliefs, and attitudes, risk communication initiators must take appropriate approaches to ensure that all social groups understand risk messages and take proactive measures to reduce and prepare for disaster hazards (Abunyewah et al. 2018).

2.4 Results

Culturally-informed risk communication is a form of transfer of risk mitigation knowledge to society, especially local communities in Indonesia (Syahputra 2019; Prawoto and Octavia 2021; Fakhriati et al. 2023; Anwar et al. 2023). The researchers try to prove this with social research claims, simplified in the form of a summary Table 2.1, of how Indonesian society uses risk communication as knowledge transfer in disaster risk reduction.

2.4.1 How Do Indonesian People Communicate, and What Forms of Communication Are Effective in Responding to Natural Hazard-Induced Disasters

Disaster preparedness is of great importance to the Indonesian people who live in disaster-prone areas. It is not surprising that they preserve knowledge with local values in the form of local wisdom, traditions, and rituals related to disasters. Before, during, and after a disaster, communication is the most effective way of responding to disasters. In the pre-disaster context, many local communities use communication in the form of oral stories passed down from generation to generation, which guide their ritual activities to prepare for various possible disasters.

The coastal communities of Lake Singkarak usually regularly carry out pilgrimage rituals to the graves of Tuanku nan Limo Puluah and Tuanku Panjang Abuak, who were believed in ancient times to be able to predict disasters with precision (Anwar et al. 2023). People often understand this when carrying out grave pilgrimage rituals while remembering local knowledge by exchanging stories from the past about signs of

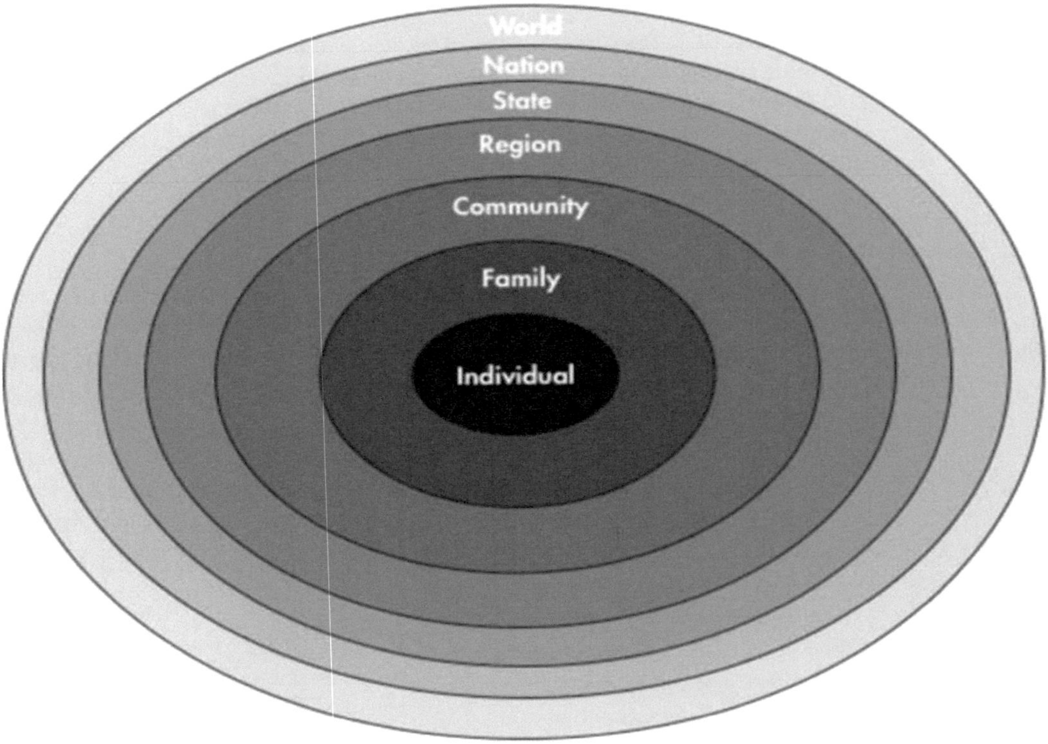

Fig. 2.1 Spheres of ethnocentricity (Sellnow et al. 2009, p. 39)

disaster and taboos for avoiding disasters and, for example, protecting forests on the shores of lakes from being destroyed to avoid floods or landslides. Additionally, knowledge inherited from ancient texts and myths is important to understand as local knowledge for disaster mitigation (Fakhriati et al. 2023; Anwar et al. 2023). It must be understood that the interaction of oral stories about tomb ecological sites, ancient texts, and myths is not just dialogue and information transfer from ordinary local communities but an effort to share common disaster mitigation cultures and accept that disasters are part of their world (Cashman and Cronin 2008; Fakhriati et al. 2023; Anwar et al. 2023).

When a disaster occurs, indigenous peoples with local knowledge passed down from generation to generation tend not to panic too much in the face of a disaster. The triangular communication pattern used by local Indonesian communities, which includes person-to-person communication based on local knowledge, has created danger signals through messages that are easy to understand (Fakhriati et al. 2023). Sundanese people shout *Jig geura narindak! Tapi ulah ngalieuk ka tukang!*" (Hurry up and run! Do not look back!) when there is an earthquake in a mountainous area to prevent people from carrying their belongings and save their lives. Residents of Simeulu, Aceh, shouted the word *Smong!* to warn the public to move immediately to save themselves from a tsunami threat (Syahputra 2019; Fakhriati et al. 2023).

Effective disaster communication allows local communities in Palu to respond after the disaster by understanding the meaning of the earthquake by developing the slogan "Palu Bangkit" and creating a liquefaction history tour in Petobo subdistrict (Allokendek and Ellisa 2022). In addition, practical media such as posters and popular books are used by the people of Kulonprogo

Table 2.1 Use of risk communication based on local knowledge culture in disaster risk reduction

No.	Journal	How do local communities (in Indonesia) communicate and prepare for natural hazard induced-disasters using traditional knowledge and practices?	How can effective disaster communication help local communities to interpret and respond to natural hazard induced-disasters and thus reduce the risk of disaster?
1.	Anwar, K., Oktavianus, O., Khanh, V. K., & Khanh, V. Q. A. (2023). Disaster mitigation through oral stories: a case study around Singkarak Lake. In E3S Web of Conferences (Vol. 464, p. 16002). EDP Sciences	Local communities communicate and prepare for earthquakes using traditional knowledge and practices through oral stories about ecological sites passed down from generation to generation	They understand the signs of natural hazard induced-disasters and build mutual understanding in a traditional process guided by religious rituals. They also maintain local wisdom in disaster mitigation by practicing disaster-related traditions and rituals, such as pilgrimages and preplanting purification rituals. This practice shows that oral storytelling is one of the traditional methods of disaster mitigation communication that is still maintained by the community
2.	Prawoto, E., & Octavia, L. (2021). Disaster storytelling: Extending the memory of the community toward disaster preparedness from myth, scientific explanation, and popular culture. Journal of Disaster Research, 16(2), 228–233	Local communities in Indonesia use indigenous knowledge and practices by incorporating traditional rituals and songs to commemorate their connection with the forces of nature. For example, the Labuan ritual is an offering to ask for safety and protection from disasters and is performed at special locations such as Mount Merapi and Parangkusumo Beach. Traditional songs such as Pangkur are also used to remind people of the dynamics of nature and the importance of building strong houses to withstand earthquakes	Combining local knowledge, rituals and traditional songs to commemorate their connection to the forces of nature. Effective disaster communication also involves collecting oral histories of disaster experiences and transforming them into practical, easily accessible media such as books and posters to ensure that knowledge is preserved and can be passed on to the next generation
3.	Syahputra, H. (2019). Indigenous knowledge representation in mitigation process: a study of communities' understandings of natural disasters in Aceh Province, Indonesia. Collection and Curation, 38(4), 94–102	The local people of Aceh pass on knowledge orally from generation to generation through traditional stories, which are only shared when the context is relevant to current conditions. They also use indigenous knowledge to understand natural behaviour and reduce the risks associated with response mechanisms and recovery alternatives based on past experience. In addition, local communities are using indigenous knowledge to develop more effective early warning systems and increase community awareness of future threats	Local communities in Aceh use indigenous knowledge to understand natural behavior, develop more effective early warning systems, and increase community awareness in facing future threats. Effective disaster communication allows governments to understand and respect local knowledge, so that they can integrate disaster risk mitigation programs from a social and cultural perspective

(continued)

Table 2.1 (continued)

No.	Journal	How do local communities (in Indonesia) communicate and prepare for natural hazard induced-disasters using traditional knowledge and practices?	How can effective disaster communication help local communities to interpret and respond to natural hazard induced-disasters and thus reduce the risk of disaster?
4.	Fakhriati, F., Nasri, D., Supriatin, Y. M., Supriadi, A., Musfeptial, M., & Kustini, K. (2023). Making peace with disaster: A study of earthquake disaster communication through manuscripts and oral traditions. Progress in Disaster Science, 18, 100287	Using indigenous knowledge and practices through what is known as triangular communication. This communication pattern involves human-to-God, human-to-human, and human-to-nature communication. This communication pattern has become part of the culture of local communities and is strengthened in the face of disasters, especially earthquakes. Local communities have developed their own ways of communicating about disasters, ranging from the dissemination of pre-disaster knowledge through ancient texts, folklore and myths, to post-disaster communication such as religious sermons to reflect on the trials they have faced	Effective disaster communication enables local communities to respond calmly to earthquakes and to understand the meaning of earthquakes from their religious and cultural perspectives. In addition, effective disaster communication enables local communities to work together to cope with earthquakes, build social solidarity and preserve the environment with vigilance. This is because they can understand and recognise the signs of disasters, respond calmly to disasters, and work together to deal with disasters
5.	Allokendek, M. L., & Ellisa, E. (2022, September). Reconditioning the resilience of Palu City within the natural disaster stories. In IOP Conference Series: Earth and Environmental Science (Vol. 1082, No. 1, p. 012020). IOP Publishing	Activities based on disaster experience are used to build solidarity and cooperation among residents in dealing with natural hazard induced-disasters, such as mutual cooperation in evacuation and post-disaster recovery. Good communication is needed between the government and the community and between the community and the community to build disaster resilience solidarity	Integrated communication between government and local knowledge helps provide accurate and timely information on early signs of natural hazard induced-disasters, preventive measures and evacuation procedures. In addition, slogans and urban history education, managed by the government and society, aim to preserve memory and pass it on to the next generation

Yogyakarta to preserve the collective memory of their elders so that local knowledge information about disasters can be understood by the current and next generations (Prawoto and Octavia 2021). This form of post-disaster communication can disseminate information that is easier to understand for local communities in the disaster area.

2.4.2 Risk Communication Based on the Local Culture, Which Must Be Sustainable

Local custom-based risk communication is still alive and well in various cultures in Indonesia. Researchers found that the coastal communities of Lake Singkarak, West Sumatra, particularly the *Kampuang Saniang Baka* (village), used the natural sign of white eagles flying around the village as a sign of an impending earthquake. It has become local knowledge, passed down from generation to generation. When someone in the community sees a white eagle, they usually inform the local religious leaders, who then pass the information on to other village communities to better prepare for disasters. Rationally, this may sound like a mythical story with no reliable scientific evidence. However, if we look more closely, we find that local knowledge like this has enough scientific basis to make animal behaviour a sign of impending disaster. Birds have better eyesight than any other vertebrate, and no bird species can match the eagle (Perrone 2009). In addition, modern molecular biology has shown that animals and birds can predict earthquakes, even 10 h before they occur (Tributsch 2005).

The people of *Kampuang Saniang Baka* are also very alert when there is no wind, and the leaves do not even move. The silence made them realize that today was not a normal silence like other days. According to their local lore, this signifies an impending earthquake. With this knowledge, they are usually prepared by leaving the house while holding the nearest board. This local knowledge is continuously passed down from previous generations through local wisdom-based risk communication from the family to the local community. This local knowledge has also been scientifically confirmed by recent research. In the study of seismic activity, the flow of the fast solar wind causes the wind to slow down upstream and form a pressure area, meaning that there are strong winds in the sky, but it looks like silence below (Anagnostopoulos et al. 2021).

The results of the researchers' interviews with informants in Palu also indicate that local knowledge is very important for the effectiveness of risk communication messages. After the 2018 earthquake, many young people began to search for the meaning of the name of the city of Palu, the meaning of the names of surrounding areas, and especially the meaning of local languages that contain disaster mitigation messages.

> ....I have lived in Palu since I was a child (27 years), but it was only after the 2018 earthquake that I found out that the meaning of the name of the city of Palu is in the regional language Topalu'e, which means raised land. Since then, I have been looking for other local languages that could be a guide to life here. I didn't know the name of my area before it became Balaroa. It was called Pusentasi. That means 'middle of the sea' in the language of the ancient people. This means that the Balaroa area was once threatened by disaster (Interview 14 November 2023. SS, Palu City.)

The results of the researchers' interviews indicate that local knowledge about disaster risk reduction in the Palu area has been around for a long time. However, risk communication with messages based on local knowledge is not perceived by the local community, especially the younger generation. According to Arifin and Zalikha (2021), *Pusentasi* is a language group of the Kaili tribe, most of whom live in the Donggala, Sigi, and Palu areas. They have local knowledge and do not want to occupy dangerous areas. For the Kaili, the Balaroa area, which they used to call *Pusentasi*, is unsuitable for settlement because it means a central hole filled with water. Since the 1980s, however, Pusentasi has changed its name to Balaroa, a busy village. Not surprisingly, the Balaroa area experienced liquefaction during the earthquake, which caused severe damage, resulting in landslides that destroyed 1357 houses (Muhanifah et al. 2021).

2.5 Discussion

Indonesia has local knowledge of disaster risk, which has been communicated within local communities. Looking at the theory used by Sellnow et al. (2009, p. 39), the ideal risk communication will be more effective if used in a smaller community, namely the local community. Risk communication can be ineffective if policymakers do not use local knowledge in disaster mitigation management. Without collaboration between the government and the local community, no messages can mobilize people's motivations to act to mitigate disasters. For example, the people of West Sumatra listen more to natural signs conveyed by religious leaders and elders based on local knowledge passed down from generation to generation. People in West Sumatra with adequate disaster literacy sometimes do not respond adequately when an earthquake occurs. They do not feel information confirmed by local leaders, such as religious leaders or elders (Fadilah et al. 2021).

In the Indonesian context, as written in Law No. 24 of 2007, social capital is one of the strengths of cooperation among disaster management stakeholders between the central government, the regional government, the National Disaster Management Agency (BNPB) and local communities (Rahmafitria et al. 2021; Koopman 2023). Social capital has included culture as trust in disaster communication sources considered part of society (Liu and Mehta 2021; Dasanayaka and Matsuda 2022). In light of this, the Indonesian government is indeed beginning to consider local knowledge as one of the risk communication messages to be used for pre-disaster mitigation and in the event of a disaster. During the 2010 Mount Merapi disaster, the government recognized the importance of balancing the trust of indigenous communities with safety measures. For example, although he remained, they worked with the Mbah Marijan to evacuate the community (Tedjokusumo 2023). Another example is the BNPB initiative to build a place of worship in the Lombok disaster camp area to bring the community together to receive messages from government officials, institutions, imams and respected parents for recovery, healing, and cohesion (Koopman 2023).

Initiatives involving local culture-based risk communication, such as the previous example, have not been fully implemented by the Indonesian government in other regions. The 2009 West Sumatra earthquake is an interesting case. The community has adequate knowledge about earthquakes, but sometimes, they do not have the motive to act in an earthquake because there is no coordination between the government and religious leaders (Fadilah et al. 2021). Furthermore, the tsunami and earthquake in Palu-Donggala caused many casualties, reportedly due to overreliance on the Tsunami Warning System (TWS). Coastal communities in these areas should be educated to interpret the natural signs of a tsunami and initiate independent evacuation when shocks are felt rather than relying solely on the Tsunami Warning System (Rafliana et al. 2022). This is consistent with research and literature findings that there is a lack of intergenerational risk communication that includes culture-based messages to explain the threat of disasters in the Palu area. The government has neglected local culture-based toponymic knowledge in regional disaster mitigation planning in the Palu area (Arifin and Zalikha 2021, pp. 25–37; Suntoro et al. 2023).

The government should synergize disaster risk communication between the government and local communities. Communicators who understand local culture are needed to strengthen local knowledge and synergize scientific knowledge from the government. Their role as communicators is very important; they can strengthen local traditions built by their predecessors through generations and raise public awareness about disaster mitigation (Fakhriati et al. 2023). Local communicators and the use of local knowledge must always be involved, as this plays an important role in mobilizing community motives for disaster reduction. This involvement can be seen in using local languages when communicating disaster messages or the involvement of traditional elders and local religious leaders as communicators (Gultom 2016; Fakhriati et al. 2023; Koopman 2023). In addition, local leaders usually have good communication skills because they have character, empathy, intelligence, leadership, and skills based on local norms (Rowan et al. 2009, pp. 168–191).

Local knowledge is important, but combining it with scientific knowledge and disseminating it to the community is much more important. Popular media, such as books and posters, are needed to reach a wider community, especially the younger generation, about local knowledge practices (Prawoto and Octavia 2021). Verified social media accounts will also integrate emergency messages and crisis management more intelligently to communities affected by disasters (Alexander 2014). Local myths and stories will also become more widely known if their truth is verified by disaster experts through social media (Hui and Tsang 2016). Governments must move with the times while maintaining local knowledge to use accurate risk communication messages.

2.6 Conclusion

Developing an ideal risk communication strategy should include recognition and respect for local culture, as local culture plays an important role in effectively disseminating information. Local knowledge has long existed in the lives of local communities to recognize the signs of disaster. The government must integrate local knowledge with science-based knowledge in disaster risk communication activities. Building trust between stakeholders and the community is the most important prerequisite for effective disaster risk communication. The government needs to recognize existing social belief networks, such as religious leaders or traditional elders, and work with them to ensure the dissemination of accurate information. Avoiding top-down methods of communicating messages through dialogue and actively responding to local community concerns will minimize local community doubts about following government instructions. By recognizing and respecting local culture in risk communication efforts, messages can be more relevant and effective in promoting understanding, mitigation, and response to potential risks.

References

Abunyewah M, Gajendran T, Maund K (2018) Conceptual framework for motivating actions towards disaster preparedness through risk communication. Proc Eng 212:246–253. https://doi.org/10.1016/j.proeng.2018.01.032

Aguirre B (2020, June 30) The myth of disaster myths. Oxford research encyclopedia of politics. Retrieved 20 Nov. 2023, from https://doi.org/10.1093/acrefore/9780190228637.001.0001/acrefore-9780190228637-e-1616

Alexander DE (2014) Social media in disaster risk reduction and crisis management. Sci Eng Ethics 20:717–733. https://doi.org/10.1007/s11948-013-9502-z

Allokendek ML, Ellisa E (2022) Reconditioning the resilience of Palu City within the natural disaster stories. IOP Conf Ser Earth Environ Sci 1082(1):012020. https://doi.org/10.1088/1755-1315/1082/1/01202

Anagnostopoulos G, Spyroglou I, Rigas A, Preka-Papadema P, Mavromichalaki H, Kiosses I (2021) The sun as a significant agent provoking earthquakes. Eur Phys J Spec Top 230:287–333. https://doi.org/10.1140/epjst/e2020-000266-2

Anwar K (2020) Ecological wisdom of oral tradition: surface and deep structure tension in preserving the lake environment. IOP Conf Ser Earth Environ Sci 469(1):012021. https://doi.org/10.1088/1755-1315/469/1/01202

Anwar K, Oktavianus O, Khanh VK, Khanh VQA (2023) Disaster mitigation through oral stories: a case study around Singkarak Lake. E3S Web Conf 464:16002. https://doi.org/10.1051/e3sconf/202346416002

Archibald MM, Ambagtsheer RC, Casey MG, Lawless M (2019) Using zoom videoconferencing for qualitative data collection: perceptions and experiences of researchers and participants. Int J Qual Methods. https://doi.org/10.1177/1609406919874596

Arifin M, Zalikha SN (2021) Pengetahuan lokal sebagai upaya pengurangan risiko bencana kota palu sulawesi tengah. In: Rahman A, Nazaruddin M (eds) Book Series Manajemen Bencana Volume 1: Pengetahuan dan Praktik Lokal untuk Pengurangan Risiko Bencana: Konsep dan Aplikasi. Syiah Kuala University Press, New York

Basher R, Ono Y (2022) Memorialization tools for systematically expanding disaster risk reduction across space and time. J Dis Res 17(4):526–531. https://doi.org/10.20965/jdr.2022.p0526

BNPB (2016) Buku Risiko Bencana Indonesia. Badan Penanggulangan Bencana Nasional: Jakarta. https://inarisk.bnpb.go.id/pdf/Buku%20RBI_Final_low.pdf

Bernard HR (2006) Research methods in anthropology: qualitative and quantitative approaches, 4th edn. AltaMira Press

Bordens KS, Abbott BB (2018) Research design and methods: a process approach, 10th edn. McGraw-Hill Education, New York

Burhani AN, Mulyani L, Pamungkas C (2021) The national research and innovation agency (BRIN): a new arrangement for research in Indonesia. ISEAS Publishing, Singapore

Burrows K, Desai MU, Pelupessy DC, Bell ML (2021) Mental wellbeing following landslides and residential displacement in Indonesia. SSM Mental Health 1:100016. https://doi.org/10.1016/j.ssmmh.2021.100016

Cashman KV, Cronin SJ (2008) Welcoming a monster to the world: myths, oral tradition, and modern societal response to volcanic disasters. J Volcanol Geoth Res 176(3):407–418. https://doi.org/10.1016/j.jvolgeores.2008.01.040

Cole TW, Fellows KL (2008) Risk communication failure: a case study of New Orleans and Hurricane Katrina. South Commun J 73(3):211–228. https://doi.org/10.1080/10417940802219702

Coombs WT (2020) Conceptualizing crisis communication. In: Heath RL, O'Hair HD (eds) Handbook of risk and crisis communication. Routledge, London, pp 99–118

Cottrell E (2015) Global distribution of active volcanoes. In: Volcanic hazards, risks and disasters. Elsevier, London, pp 1–16. https://doi.org/10.1016/B978-0-12-396453-3.00001-0

da Cunha C, Rocha APF, Cardon M, Breton F, Labeyrie L, Vanderlinden JP (2020) Adaptation planning in France: inputs from narratives of change in support of a community-led foresight process. Clim Risk Manag 30:100243. https://doi.org/10.1016/j.crm.2020.100243

Dasanayaka U, Matsuda Y (2022) Role of social capital in local knowledge evolution and transfer in a network of rural communities coping with landslide disasters in Sri Lanka. Int J Dis Risk Reduct 67:102630. https://doi.org/10.1016/j.ijdrr.2021.102630

Denzin NK, Lincoln YS (eds) (2018) The SAGE handbook of qualitative research, 5th edn. SAGE Publications

Fadilah M, Maryani E, Permanasari A, Riandi R (2021) Disaster-vulnerable community perception related to pre-earthquake natural phenomena in west Sumatera as part of disaster preparedness. IOP Conf Ser Earth Environ Sci 683(1):012075. https://doi.org/10.1088/1755-1315/683/1/012075

Fakhriati F, Nasri D, Supriatin YM, Supriadi A, Musfeptial M, Kustini K (2023) Making peace with disaster: a study of earthquake disaster communication through manuscripts and oral traditions. Prog Dis Sci 18:100287. https://doi.org/10.1016/j.pdisas.2023.100287

Fuady M, Munadi R, Fuady MAK (2021, February) Disaster mitigation in Indonesia: between plans and reality. In: IOP conference series: materials science and engineering (Vol 1087, No 1, pp 012011). IOP Publishing. https://doi.org/10.1088/1757-899X/1087/1/012011

Gerring J (2016) Case study research: principles and practices. Strategies for social inquiry, 2nd edn. Cambridge University Press, Cambridge. https://doi.org/10.1017/9781316848593

Gleckler PJ, AchutaRao K, Gregory JM, Santer BD, Taylor KE, Wigley TML (2006) Krakatoa lives: the effect of volcanic eruptions on ocean heat content and thermal expansion. Geophys Res Lett 33:L17702. https://doi.org/10.1029/2006GL026771

Grabill JT, Simmons WM (1998) Toward a critical rhetoric of risk communication: producing citizens and the role of technical communicators. Tech Commun Q 7(4):415–441. https://doi.org/10.1080/10572259809364640

Groot G, Waldron T, Barreno L, Cochran D, Carr T (2020) Trust and world view in shared decision making with indigenous patients: a realist synthesis. J Eval Clin Pract 26(2):503–514. https://doi.org/10.1111/jep.13307

Gultom DI (2016) Trustworthy and participatory community-based disaster communication: A case study of Jalin Merapi in the 2010 Merapi eruption in Indonesia (Doctoral dissertation, University of Canterbury). University of Canterbury Research Repository. https://ir.canterbury.ac.nz/server/api/core/bitstreams/12656cf3-0866-4591-b85e-5f06d239efa1/content

Gupta M, Sukamto KE (2020) Cultural communicative styles: the case of India and Indonesia. Int J Soc Cult Lang 8(2):105–120

Heath RL, O'Hair HD (eds) (2020) Handbook of risk and crisis communication. Routledge, New York

Hilburn AM, Zawisza TT (2019) River of difference: using participatory risk mapping to assess perceived risks in Laredo, Texas, USA, and Nuevo Laredo, Tamaulipas, Mexico. In: Risk communication and community resilience. Routledge, New York, pp 69–85. https://doi.org/10.4324/9781315110042

Hui LHD, Tsang PKE (2016) Everyday knowledge and disaster management: the role of social media. In: Robertson M, Tsang P (eds) Everyday knowledge, education and sustainable futures. Education in the Asia-Pacific region: issues, concerns and prospects, vol 30. Springer, Singapore. https://doi.org/10.1007/978-981-10-0216-8_8

Juneza CI, Putri SMR, Novella U, Kakkimoto K, Ueno Y, Fujiyama I (2023) Japan and Indonesia comparative study: changes in communication and impact on the level of social presence of university student organizations before and during the pandemic. E3S Web Conf. https://doi.org/10.1051/e3sconf/202338804052

Kar B, Cochran DM (eds) (2019) Risk communication and community resilience. Routledge, London

Kinoshita T (2014) Short history of risk communication in Japan. J Dis Res 9(1):592–597. https://doi.org/10.20965/jdr.2014.p0592

Koopman J (2023) Subawe, traditional knowledge, and faith-based organisations promoting social capital and disaster preparedness: a Lombok, Indonesia case study. Int J Dis Risk Reduct 94:103837. https://doi.org/10.1016/j.ijdrr.2023.103837

Kurnio H, Fekete A, Naz F, Norf C, Jüpner R (2021) Resilience learning and indigenous knowledge of earthquake risk in Indonesia. Int J Dis Risk Reduct 62:102423. https://doi.org/10.1016/j.ijdrr.2021.102423

Laurila-Pant M, Pihlajamäki M, Lanki A, Lehikoinen A (2023) A protocol for analysing the role of shared situational awareness and decision-making in cooperative disaster simulations. Int J Dis Risk Reduct 86:103544. https://doi.org/10.1016/j.ijdrr.2023.103544

Lin KHE, Khan S, Acosta L, Alaniz R, Olanya D (2020) The dynamism of post disaster risk communication: a cross-country synthesis. Int J Dis Risk Reduct. https://doi.org/10.1016/j.ijdrr.2020.101556

Littlefield RS, Cowden K (2006, November) Rethinking the single spokesperson model of crisis communication: recognizing the need to address multiple publics. In: Annual convention of the national communication association, San Antonio, Texas. Accessed via the link. https://www.researchgate.net/publication/228418318

Liu BF, Mehta AM (2021) From the periphery and toward a centralized model for trust in government risk and disaster communication. J Risk Res 24(7):853–869. https://doi.org/10.1080/13669877.2020.1773516

Lobe B, Morgan D, Hoffman KA (2020) Qualitative data collection in an era of social distancing. Int J Qual Methods. https://doi.org/10.1177/1609406920937875

Marchezini V, Trajber R, Olivato D et al (2017) Participatory early warning systems: youth, citizen science, and intergenerational dialogues on disaster risk reduction in Brazil. Int J Dis Risk Sci 8:390–401. https://doi.org/10.1007/s13753-017-0150-9

McIntosh I, Wright S (2019) Exploring what the notion of 'lived experience' offers for social policy analysis. J Soc Policy 48(3):449–467. https://doi.org/10.1017/S0047279418000570

Mercer J, Kelman I, Taranis L, Suchet-Pearson S (2010) Framework for integrating indigenous and scientific knowledge for disaster risk reduction. Disasters 34(1):214–239. https://doi.org/10.1111/j.1467-7717.2009.01126.x

Muhanifah H, Adi AD, Faris F (2021) Liquefaction investigation of Balaroa, Central Sulawesi on liquefied and non-liquefied areas. IOP Conf Ser Earth Environ Sci 861(5):052039. https://doi.org/10.1088/1755-1315/861/5/052039

Nagamatsu S, Fukasawa Y, Kobayashi I (2021) Why does disaster storytelling matter for a resilient society? J Dis Res 16(2):127–134. https://doi.org/10.20965/jdr.2021.p0127

Nishimura S, Nevgi A, Tella S (2008) Communication style and cultural features in high/low context communication cultures: A case study of Finland, Japan and India. Teoksessa A. Kallioniemi (toim.), Uudistuva ja kehittyvä ainedidaktiikka. Ainedidaktinen Symposiumi 8(2008):783–796

Palenchar MJ, Heath RL (2002) Another part of the risk communication model: analysis of communication processes and message content. J Public Relat Res 14(2):127–158. https://doi.org/10.1207/S1532754XJPRR1402_3

Palmlund I (2020) Risk and social dramaturgy. In: Heath RL, O'Hair HD (eds) Handbook of risk and crisis communication. Routledge, London, pp 192–204

Perrone NR (2009) Ho the heightened senses of the sea Turtle, Elephant, and Eagle Are E ident in the Brain. Yale-New Haven Teachers Institute, Volume IV

Plough A, Krimsky S (1987) The emergence of risk communication studies: social and political context. Sci Technol Hum Values 12(3/4):4–10

Prawoto E, Octavia L (2021) Disaster storytelling: extending the memory of the community toward disaster preparedness from myth, scientific explanation, and popular culture. J Dis Res 16(2):228–233. https://doi.org/10.20965/jdr.2021.p0228

Pribadi KS, Abduh M, Wirahadikusumah RD, Hanifa NR, Irsyam M, Kusumaningrum P, Puri E (2021) Learning from past earthquake disasters: the need for knowledge management system to enhance infrastructure resilience in Indonesia. Int J Disaster Risk Reduct 64:102424. https://doi.org/10.1016/j.ijdrr.2021.102424

Purnomo A, Kurniawan B (2019) Dhamar Wulan and Menak Jinggo War; Mythological Interpretation on The Perception of Disaster Risk at Raung Eruption. IOP Conf Ser Earth Environ Sci 273(1):012031. https://doi.org/10.1088/1755-1315/273/1/012031

Rafliana I, Jalayer F, Cerase A, Cugliari L, Baiguera M, Salmanidou D, Hancilar U (2022) Tsunami risk communication and management: contemporary gaps and challenges. Int J Dis Risk Reduct 70:102771. https://doi.org/10.1016/j.ijdrr.2021.102771

Rahmafitria F, Sukmayadi V, Suryadi K, Rosyidie A (2021) Disaster management in Indonesian tourist destinations: how institutional roles and community resilience are mediated. Worldw Hosp Tour Themes 13(3):324–339. https://doi.org/10.1108/WHATT-01-2021-0014

Rahman A, Nazaruddin M (eds) (2021) Book Series Manajemen Bencana Volume 1: Pengetahuan dan Praktik Lokal untuk Pengurangan Risiko Bencana: Konsep dan Aplikasi. Syiah Kuala University Press, London

Rampino MR, Self S (1982) Historic eruptions of Tambora (1815), Krakatau (1883), and Agung (1963), their stratospheric aerosols, and climatic impact. Quatern Res 18(2):127–143. https://doi.org/10.1016/0033-5894(82)90065-5

Rowan KE, Botan CH, Kreps GL, Samoilenko S, Farnsworth K (2009) Risk communication education for local emergency managers: using the CAUSE model for research, education, and outreach. In: Handbook of risk and crisis communication. Routledge, London, pp 168–191

Schwartz-Marin E, Merli C, Rachmawati L, Horwell CJ, Nugroho F (2022) Merapi multiple: protection around Yogyakarta's celebrity volcano through masks, dreams, and seismographs. Hist Anthropol 33(5):588–610. https://doi.org/10.1080/02757206.2020.1799788

Sellnow TL, Ulmer RR, Seeger MW, Littlefield RS (2009) Introducing a message-centered approach to risk communication. In: Effective risk communication. food microbiology and food safety. Springer, New York. https://doi.org/10.1007/978-0-387-79727-4_1

Septiana ME, Wardoyo MAI, Praptiwi NY, Ashari ANS, Ashari A, Susanti NI, Nugrahagung PP (2019, June) Disaster education through local knowledge in some area of merapi volcano. In: IOP conference series: earth and environmental science (Vol 271, No 1, pp 012011). IOP Publishing. https://doi.org/10.1088/1755-1315/271/1/012011

Shafie AAHB, Othman K, Baharudin DF, Najaa-Mokhtar A, Rahimi MKA, Jaladin RAM, Kama NFM (2022) A review on online focus group discussion during the covid-19 pandemic. J Pharmaceut Neg Results. https://doi.org/10.47750/pnr.2022.13.S09.692

Snyder H (2019) Literature review as a research methodology: an overview and guidelines. J Bus Res 104:333–339. https://doi.org/10.1016/j.jbusres.2019.07.039

Suntoro A, Wahyuni T, Ismeti, Utomo NA (2023) Toponym and Kayori: potential benefits of local wisdom to Kaili in disaster mitigation strategies in central Sulawesi, Indonesia. In: Indigenous knowledge and disaster risk reduction: insight towards perception, response, adaptation and sustainability. Springer, Cham, pp 321–342. https://doi.org/10.1007/978-3-031-26143-5_15

Syahputra H (2019) Indigenous knowledge representation in mitigation process: a study of communities' understandings of natural disasters in Aceh Province, Indonesia. Collect Curat 38(4):94–102. https://doi.org/10.1108/CC-11-2017-0046

Tanhas IMWA, Hayati Y, Nst MI (2017) Perbandingan Struktur Cerita Rakyat Legenda Bujang Sambilan di Nagari Paninggahan Kecamatan Junjung Sirih Kabupaten Solok dengan Struktur Cerita Rakyat Legenda Bujang Sambilan di Pasa Akaik Kecamatan Tanjung Raya Kabupaten Agam. Jurnal Bahasa dan Sastra 5(1):110–123. https://doi.org/10.24036/898670

Tedjokusumo Z (2023) Indigenous ways of knowing for disaster preparedness in Indonesia. Aalto University, Berlin

Trainor A, Graue E (eds) (2018) Reviewing qualitative research in the social sciences. Routledge, New York

Tributsch H (2005) The bionic anticipation of natural disasters. J Bionic Eng 2(3):123–144. https://doi.org/10.1007/BF03399489

Vagle MD (2018) Crafting phenomenological research, 2nd edn. Taylor and Francis, London. https://doi.org/10.4324/9781315173474

van Jankowski NW, Selm M (2005) Epilogue: methodological concerns and innovations in internet research, p 207. https://repository.ubn.ru.nl/handle/2066/56846

World Health Organization (2020) Risk communication and community engagement readiness and response to coronavirus disease (COVID-19): interim guidance, 19 March 2020. World Health Organization. https://apps.who.int/iris/handle/10665/331513. License: CC BY-NC-SA 3.0 IGO

3 Considering Risk Culture to Strengthen Disaster Risk Governance in Indonesia

Gusti Ayu Ketut Surtiari and Deny Hidayati

Abstract

This study examines the current state of Indonesia's risk governance when facing more frequent and intense disasters with uncertain impacts on social and ecological systems. Indonesia, as a high-risk country, has progressively committed to reducing the risk of disasters since the disaster management law was enacted in 2007, followed by the establishment of the National Agency for Disaster Management (BNPB). The transformative pathway can be seen in the country's report on climate change adaptation based on the Paris Agreement 2015 and disaster risk reduction based on the Sendai Framework 2015–2030. However, the challenges persist in treating a risk-awareness ecosystem among various stakeholders that showed organically during disaster lack of preparedness and coordination to respond to ineffective responses during emergency periods. We argue that risk culture is important for more effective risk governance because it involves local values and beliefs. This study is based on literature reviews that are mainly based on the authors' research reports in Indonesia, which consist of disaster preparedness and recovery. Using the Systemic Risk Governance approach for the Indonesian context, this study explores the potential consideration of risk culture in Indonesian risk governance. The results show that historically, Indonesia has established mitigation strategies for facing natural hazards. However, the uncertainty and complexity of the current social-ecological system have caused difficulties in its integration. This study is useful in finding an effective strategy to strengthen risk governance.

Keywords

Risk governance · Disaster risk reduction · Climate change adaptation · Systemic risk governance · Risk culture

3.1 Introduction

The increased risk of climate, geological, and biological hazards, followed by increased uncertainty in the future, has forced vulnerable countries such as Indonesia to strengthen their risk governance. Risk governance promotes a more systemic approach to managing risk and uncertainty by identifying, assessing, and mitigating individuals, organizations, and communities (Renn 2008). It emphasizes the complex system of actors, rules, conventions, processes,

G. A. K. Surtiari (✉) · D. Hidayati
Research Center for Population, BRIN, Jakarta, Indonesia
e-mail: gust003@brin.go.id

D. Hidayati et al. (eds.), *Examining Disaster Risk Reduction in Indonesia*,
Sustainable Development Goals Series, https://doi.org/10.1007/978-981-96-4182-6_3

and mechanisms, considering contextual factors, such as institutional arrangements, by differentiating vertical and horizontal governance (Renn 2012). The Sendai Framework 2015–2030 also highlighted the importance of managing the multi-hazards, multi-temporal, multi-dimensions of vulnerability and residual risks through disaster risk governance (Triyanti et al. 2023). The current stage of risk governance worldwide is diverse, depending on the capacity to build resilience across levels and effectively reduce the risk of disasters.

Indonesia is one of the most vulnerable countries to climate-related and natural hazards that severely affect the social and ecological system in Indonesia (Bnpb 2013; Amri et al. 2016). The number of disasters has increased substantially with the devastating losses of social, financial, and environmental assets and life. Earthquakes, tsunamis, and volcano eruptions devastate society and the economy, with hundreds of thousands of lives lost to these disasters. More than 170,000 people lost their lives because of the earthquake and tsunami in Aceh, causing more than 37,000 people to go missing and 500,000 to be displaced.

Frequent earthquakes and tsunamis followed, killing about 200 people in Nias (March 2005) and more than 5000 people in Yogyakarta (March 006). Between 2017 and October 2018, approximately 11,352 people died, and 33.5 million people were displaced by 20,342 disaster events (BNPB 2014, 2018). The recent severe earthquake and tsunami in Palu have caused more than 2000 people dead and around 18 trillion in economic losses. Meanwhile, the impact of climate change, such as sea level rise and more frequent and intense extreme floods, threatens food security and deepens the vulnerability of vulnerable populations.

Since the Aceh tsunami, the Indonesian population developed a variety of preparedness activities, including various capacity building among the government's staff. However, the significant number of lives lost during disasters and significant economic loss indicates that the disaster preparedness is still lacking as a result of their lack of knowledge and awareness (Hidayati 2012). The knowledge and awareness of the community and government have increased understanding to improve their preparedness and take the action needed to anticipate earthquakes and tsunamis.

There is still a lack of preparedness for emergency planning, such as evacuation maps, places and routes, first aid and evacuation skills and drills, and an emergency kit. In addition, there is a limited capacity to access information about early warnings and to respond to warnings, increasing resource mobilization capacity, including knowledge and skills to save lives and reduce disaster risks. The condition mentioned earlier is closely related to the fact of absence of a risk culture (Cornia et al. 2016). Studies show that attitudes, behaviors, values, and practices are components of culture that effectively influence better risk governance (Tong-Fu et al. 2023; Lanza et al. 2023; Marshall 2020).

This study examines the current state of Indonesia's risk governance when facing more frequent and intense disasters with uncertain impacts on social and ecological systems. We use the approach of Systemic Risk Governance (SRG) for Indonesia that has been reviewed by Triyanti et al. (2023). The importance of using the frame of SRG is regarding the complexity of risk governance particularly in the Indonesian context. Moreover, the using of the SRG lens also allows us to examine risk governance from a wider perspective, not only from an economic perspective but also root causes of real risk (Ibid.). The SRG is an approach that could be used for further understanding of specific governance entities such as actors, institutions, and mechanisms that will determine the management of risks IRGC 2018).

3.2 Theoretical Framework

3.2.1 Approaching Risk Governance from the Perspective of the Integration of Disaster Risk Reduction and Climate Change

The increased uncertainty and complexity of global phenomena require a new perspective to manage risk. Disasters to natural and climate-related hazards are growing to challenge the success of development goals. Therefore, managing risk and development needs to be more integrated particularly to adapt to climate change impact and reduce risk to disaster. The notion of risk governance was promoted in 2001 as a new paradigm to govern risk in a modern perspective. It highlights the decision-making process which considers all the processes of identification of the problem, assessment, management, and communication (van Asselt and Renn 2011). The risk assessment and risk management should be communicated in the process of decision-making at all levels and scales. This consideration will be able to avoid unintended consequences within risk management and effort to build resilience. Many studies find that strategy to adapt and risk reduction of one group at particular level an indifferent time could trigger another risk for another actors or institutionalization of risk management could limit local capacity to adapt and manage their risk (O'Brien and Leichenko 2000; Birkmann 2011; Eriksen et al. 2011).

The notion of risk governance is gaining its importance, as stated in the Sendai Framework. The Sendai Framework addresses strengthening disaster risk governance as the priority two after the priority one, to understand disaster risk, followed by investing disaster risk reduction for resilience as priority three and enhancing disaster preparedness for a build back better in the priorities (UNISDR 2015). As a continuation of Hyogo framework, Sendai Framework also highlights the link between and development which is even relevant for the target to implement Paris Agreement.

The Paris Agreement in 2015 articulates the commitments of countries to respond climate change by putting target to keep global temperature in the level of 1.5 degrees Celsius. The risk of natural hazards could threaten the success of development and or development can create risk when it neglects the possibility of uncertainty (UNISDR 2015). For instance, landslides in the mountainous areas are only potential risk to climate-related hazards, but when it is the core of human activities areas for residential and economic activities thus it is becoming a risk to disaster. When landslides occur, it will trigger severe loss and damage because the number of people exposed by landslide and related infrastructures built in those areas. The scheme linking risk and development is part of risk governance (Renn 2008).

Risk governance acknowledges the complexity of a system within society and the linkage with the environment. Risk management itself will consider the multi-aspect with multi-stakeholders at all levels, and various scales and those have interconnected with each other (Renn 2008; Mauelshagen et al. 2014). Risk governance could also be explored from multiple approaches, various actors, and across scales and levels (Cedergren and Tehler 2014). The traditional risk management recently focuses on the handling risk in a very narrow context (Cedergren and Tehler 2014). For instance, risk management on earthquake and tsunami but still lacking to consider risk caused by infrastructures that affected by hazards. The recent disaster shows several cascading extreme events floods caused by heavy rain and the leaking of dike and reservoir that caused devastating inflow of water to the people. The disaster caused by multi-hazards is getting more frequent. It influences the capacity to cope and anticipate among vulnerable people.

Risk governance is a transformation of substance and principal values of governance in the context of risk-related decision-making (van Asselt and Renn 2011). The terms of governance designate a multitude of actors and process toward the necessary collective decision (Ibid.). Multilevel stakeholders comprise diverse actors, ranging from national to multinational entities, governmental to non-governmental organizations, private to public bodies, and individuals to communities. These stakeholders navigate uncertain, complex, and ambiguous environments, and they are all interconnected (Pahl-Wostl et al. 2013). Moreover, in the risk governance approach, the aspect of risk will be more integrated and consider the cascading risk across scales and level (Renn 1998). The decision-making process will not only focus on one aspect of the risks, only the people, but it also consider risk caused by infrastructures, social, economic, and environment. It involves many actors, institutions, public, and privates and also handles uncertainty and ambiguity.

It is important to note that risk governance process will allow the various stakeholders to set their activities and take actions to manage the uncertainty and complexity (Djalante et al. 2011; Bronen and Chapin 2013; Birkmann et al. 2014). Each stakeholder will consider participation and collaboration that consist of leadership, trust and social capital, self-organization or network, and learning and innovation both public social learning and institutional memories (Djalante et al. 2013). Effective stakeholder engagement necessitates the direct interplay of all these components. For instance, leadership is needed in the multilayered institutions and organizations involved in disaster management. Moreover, leadership also influences the ability to self-organize a society (Ibid.). Learning and innovation will be optimal in the case where a system can self-organize, and there are several actors involved and collaborate to manage risk to disaster.

The evaluation of risk governance in Indonesia was started in 2005 as an initial year of implementation of the Hyogo Framework. The progress result has been reported and published. It shows that the role of leadership is important, but at the local level, awareness and knowledge on risk among the local leader is still low. This might impact the planning and implementation of disaster management. Fundamentally, the challenges to risk governance in Indonesia extend to systemic deficiencies in technology, infrastructure, and institutional capacity. The case of Merapi Eruption shows explicitly the problem of lack of infrastructures, complex interaction across institutions, dependence on funds from external parties, and limited documentation on human and material loss (Bakkour et al. 2015).

3.2.2 Highlight the Importance of Risk Culture to Strengthen Risk Governance

Risk culture is a terminology that describes the values, beliefs, and knowledge that shape the attitudes and behavior about the risks faced and shared by a community or group with a common objective. It is a social construction developed by a group of people or organizations for a long time through a nexus of meaning and assumptions. Social structures influence a community or group to define, determine causes, convey moral judgments, and recommend improvements. Risk culture should be analyzed contextually by applying the eventuality and specific approach to deal with risks and disasters (Renn 2008; Cornia et al. 2014).

Other references state that risk culture is a general expression to show facts that risk perception and disaster response are strongly influenced by socio-cultural factors. It is also an intrinsic feature of a community where people increasingly perceive risks caused by themselves (from the development of science and technology). Risk culture links different types of risks (about frequency and severity) and various stakeholders. The pattern of risk perception and management of disaster, therefore, varies among communities and nations in time and space (Schmit and Roth 1990; Zwick and Renn 2002; Wachinger et al. 2013).

Cornia et al. (2014) pointed out three types of risk culture: an individual-oriented risk culture, a state-oriented risk culture, and a fatalistic risk culture. Each model is strongly influenced by disaster framing, trust in authorities, and responsibility attribution for risk prevention. Individual-oriented risk culture focuses on self-reliance and individual responsibility. In an individual-oriented risk culture, individuals are broadly dependent on themselves in mastering disasters.

They believe that they could prevent and minimize risks through the adoption of appropriate methods.

People are intended to be aware of the risks where they live, to check about their housing construction based on standard building codes, to pay attention to early warning from relevant authorities, and to follow insurance to reduce the impacts of disaster. They become actors in reducing risk and increasing disaster mitigation and preparedness. They have high-risk awareness, knowledge, and coping strategies, and these become their capacity to face critical conditions. People feel that they have responsibility for their security, although they understand that the state has the responsibility to deliver information about disaster risk and provide safety of its people during and after the crises.

A state-oriented risk culture largely relies on disaster prevention. Disaster risk reduction and management have become the state responsibility, mainly through its emergency management institution. In contrast to individual-oriented risk culture, in state-oriented risk culture people rely heavily on the state. The state is considered as a 'paternalistic' who must protect the safety of its people. High dependence of people on the state inhibits their risk awareness and self-reliant solutions, mainly because of their profound knowledge and coping strategies. The fatalist risk culture is based on people's perception that they are unprotected and powerless vis a vis disaster. Hazards are framed as exceptional events or external factors and perceived as unpredictable and unavoidable phenomena. There is a perception that neither the state nor its people possess the full capacity to prevent risks, manage critical situations, or adequately mitigate disaster impacts. Risk prevention has therefore not become a crucial concept in this risk culture. This leads to low-risk awareness, knowledge, and coping strategies.

Risk culture is a decent indicator to explain how disaster risk management and practice have been implemented. It covers the community's and government's awareness, attitudes, and behaviors toward risk. The community and government, therefore, must first understand the existence of risk culture and gauge how good it contributes to the community's and institution's risk strategy and management method. It will be useful when the community and government understand their approach to risks, take responsibility to manage risks, and encourage others to follow their example. The risk culture, therefore, cannot be a stand-alone component, but it should be linked to risk governance practices and incentive programs to achieve effective risk management.

How cultures deal with risks and disasters can be explained through disaster framing, blaming, and trust in authorities (Cornia et al. 2014). It describes how communities, groups and all institutions (including private companies, government and non-government bodies) share common ways in perceiving risk, common knowledge about how to handle disaster, common beliefs about who should be accused for the disaster consequences, everyday feelings of trust or mistrust toward authorities and appropriate information to be adopted in case of emergencies. Renn (2008) also stated that risk perception should be placed together with risk communication and risk management within risk governance. Risk perception is an individual or subjective perspective but it represents the beliefs, attitudes, and values at the local communities (Wachinger et al. 2013).

Cornia et al. (2014) classified the relationship between risk and culture into three typologies: (1) Cultural dominance in influencing the community to deal with threats and disasters. This typology is based on the close relation between a society with the environment around, including its natural hazards, it, therefore, varies from one group to others; (2) Community perception about risk and disaster, how the community explains risks and disasters and focuses on risk awareness, accordance to authorities and communication. This typology tends to simplify and generalize the process because it focuses on the dominant trends characterizing a group; and (3) Responsible stakeholders for disaster consequences, authorities, or culture-based nations owned by community members.

As part of risk culture, studies find that local wisdom plays a vital role in shaping risk culture through the perception of the people and the way they communicate risk. Moreover, local knowledge could play a critical role in strengthening community disaster preparedness for effective response and its potential for building back after recovery. It can also be applied to strengthen community efforts to find relief and recover from the impact of the disaster. Knowledge, values, and action can be developed in harmony to enhance individual or group's capacity to reduce disaster risk.

3.2.3 The Systemic Risk Governance: A Potential Approach to Facilitate Involving Risk Culture

Departing from the guidelines for The Governance of Systemic Risk (SRG) developed by IRGC (2018), the processes consist of seven steps. (1). Exploring the system by defining the boundaries and dynamics of the system. (2). Developing scenarios of the risks. (3). Determining the objectives and achievements including the level of tolerability of risk and uncertainty. (4). Co-developing management strategies. (5). Addressing unanticipated barriers and sudden critical shifts, (6). Deciding, testing, and implementing the strategies. (7) Monitoring, learning from the strategy implementation followed by reviewing and adaptation. Implementation in the case of Indonesia, Triyanti et al. (2023) observed deeply the six steps to go forward with systemic risk governance namely, (1) social scanning to have local context overview, (2) development of scenario for multiple time horizons, (3) determination of risk tolerability based on expert judgment, (4) considering the local context for the entire process of the decision-making process, (5) successful of strategies should consider cognitive aspect, and (6) involving social justice, transparency, accountability, and appropriateness in risk governance.

Studies on risk governance are closely related to the concept of resilience since risk governance represents a decision-making process with multiple actors, across scales and levels. Resilience is a system dynamic with multiple systems, transitions, and thresholds with cross-scale interaction, cascading change, and highlighting system governance of social networks and adaptive governance (Resilience Alliance 2010). A resilience approach is needed for risk governance since its approach undertakes the uncertainty and complexity of the system and the link between social and ecological systems (Ibid.). The connection across scales can be described as the process of a system absorbing shocks and empowering resources to transform themselves to be more resilient instead of being vulnerable. Resilience is defined as the ability of a system, community, or society exposed to hazards to resist, absorb, accommodate, and recover from the effects of a hazard in a timely and efficient manner, including through the preservation and restoration of its essential basic structures and functions (UNISDR 2009).

3.3 Methodology

3.3.1 Tools to Examine Risk Governance in Indonesia to Incorporate Risk Culture

This study aims to examine the roles of risk culture to strengthen risk governance in Indonesia. Considering the advanced progress of risk management in Indonesia, we apply the SRG framework to explore the potential to involve risk culture within risk governance. By unveiling the complexity of governing risk, we could prioritize at which stage the intervention to effectively implement an approach to reduce the risk of disaster.

The scoping study of this research is understanding the potential and challenges of involving risk culture in Indonesian risk governance. As we argue that the progressive efforts of the Indonesian government to achieve the target of successful adaptation as addressed in the Paris Agreement and disaster risk reduction as guided through the Sendai Framework, inclusivity is needed. We use risk culture as one of the components of the inclusive approach to effectively implement risk governance. The six stages as proposed by Triyanti et al

(2023) will be used as guidance to determine in which stage of risk culture it is possible to be involved in and show its effectivity.

3.3.2 Desk Review of Four Major Disasters

This study examines some major disasters in Indonesia such as the Tsunami in Mentawai and Semeleu, the Earthquake in DI Yogyakarta and Lombok, the Earthquake in Padang, West Sumatera, the Tsunami in Palu, and some other climate-related hazards such as floods and drought. The selection of the disaster events is based on the active involvement of authors on the research project and also based on respective published research papers to complement the review (Table 3.1).

Table 3.1 Major disasters in Indonesia and local wisdom

No.	Disaster (large scale)	Local wisdom (culture aspect)
1.	The Indian Ocean tsunami, 2004 (Aceh)	Warning system through local wisdom (Smong)
2.	A typical tsunami in Central Sulawesi, 2018	Toponomi
3.	Earthquake in Yogyakarta, 2005	Mutual assistance (social capital of gotong royong)
4.	Earthquake in Lombok, 2018	Traditional housing that is resilient to earthquake from Sasak community namely Bale Tani
5.	Earthquake in Padang, 2009	The role of religious leaders for warning system

3.4 The Current and Future Progress of Risk Governance in Indonesia: Learning from Major Disasters

3.4.1 Overview of Progress on Risk Governance in Indonesia

In 2007, Indonesia had a Law 24/2007 on Disaster Management that followed by laws and regulations to support the implementation of that law. For instance, Law 26/2007 on the spatial plan and the establishment of BNPB. BNPB is the core actor with the function of coordinating all related stakeholders to implement disaster management in Indonesia. After BNPB had been established, the Local Agency for Disaster Management (*Badan Penanggulangan Bencana Daerah*/BPBD) also had to be installed at the provinces and regency or city levels. In 2015, there is around 90 percent of BPBD was built in all cities and regencies' level. BPBD is the main actor at the local level for disseminating and implementing the DRR agenda together with the Local Agency for Development Planning (BAPPEDA). However, BPBD is ruled by two regulations namely the Regulation of Minister of Home Affair 46/2008 and the Regulation of BNPB 3/2008. This means, BPBD is not entirely responsible for BNPB but also depends on the local government coordination.

The Law 24/2007 addresses the shared responsibilities of disaster management in Indonesia. It is not only one agency or national body, but also it must be shared with governmental organizations at multiple scales. Non-government organizations also have substantial roles in supporting the implementation of Law 24/2006 at the local and community level. Several NGOs contribute to the effort of building resilience at the community level. Most of them are donor agencies, the United Nations, and the Red Cross Societies (Djalante 2013). In Jakarta, for instance, NGOs complement the absence of government to support the most vulnerable people in slums and informal settlements (Surtiari et al. 2019). There are international organizations with local collaboration to empower and disseminate knowledge and awareness on the risk of disaster and to increase capacity to build resilience. In Jakarta for instance, the Wahana Visi as part of the World Vision organization in the local community, OXFAM and Save the Children closely collaborate with local communities to empower slum dwellers to have a better capacity to cope with extreme floods. Also, various local NGOs intensively support the building resilience of vulnerable people by disseminating local early warning systems and knowledge on the risk of wave floods.

While the institutionalization of disaster management and climate change adaptation has advanced, in the implementation stage, there remain several problems. Earthquakes and tsunamis in Palu have shocked all scientists and practitioners regarding the scale of the impact of disasters. Palu is one of the cities in Indonesia that developed a disaster contingency plan in 2012. The contingency plan clearly stated the risk of earthquakes and tsunamis and the information for early warning and evacuation routes. The city major of Palu also received international recognition in terms of its commitment to disaster preparedness. After the contingency plan was enacted, in the same year, the disaster preparedness group (*Kelompok Siaga Bencana*/KSB) was formed to facilitate the Palu tsunami drill led by BNPB (Rafliana 2018). The KSB has been drilled to do evacuation before the national drill. The national drill involved the potentially affected communities, but our study finds that most people only watched the tsunami drill instead of fully understanding of the evacuation. They could mention there are several stakeholders who joined the tsunami drill.

However, the study conducted by LIPI (Surtiari et al. 2019) shows that none of the local government agencies were aware of the document except certain people in BPBD. At the community level, only a few people could remember the tsunami drill and some of them only watched the simulation instead of joining it. Therefore, the evacuation route and shelter or meeting points are not well recognized by the population and some of the local government. During the tsunami in 2018, the affected community evacuated themselves spontaneously and followed others. They are only aware that they have to go to a higher place but they have no information about what should they do during the evacuation. The dissemination of evacuation among students was conducted but only once and in a very short time. This process is complicated to be understood and memorized after dissemination.

The finding shows the constraint to implement the contingency plan prompted by the limited budget for the maintenance of the program. The city of Palu prefers to prioritize routine development programs instead of focusing on disaster preparedness. The responsible local agencies find it difficult to allocate their resources to implement the contingency plan due to less capacity and knowledge on disaster risk reduction. However, during the emergency period, all related local agencies intensively play their roles and collaboratively act to manage the emergency.

3.4.2 Inventory of Local Wisdom to Respond Natural Hazards

3.4.2.1 Local Knowledge of Hazardous-Prone Areas Through Toponyms

The tsunami in Palu and its surroundings in 2018 was not the first tsunami to occur in the region. A major earthquake followed by a tsunami occurred in 1927 and then occurred again in 1938 (Triyanti et al. 2023). At the community level, people are aware of the risk of earthquakes and tsunamis. The people can mention some of the earthquakes in previous years. However, it is only on their knowledge without any consciousness to make a specific decision to reduce risk or avoid risk. Nevertheless, people also could mention the history of their land as such, and the coastal areas reach the hilly areas where they are currently building their homes. That knowledge is unable for people to learn to adapt and take robust action to manage risk. Only one community who lives in Labena, Sirenja, and Batusuya villages could save their lives because they do self-evacuation. After the first shaking, the Hamlet leader instructed the people to evacuate to the hilly place by bringing some clothes and a ripped tent (Rafliana 2018). They learned from the earthquake in 1968, and they still remember how to evacuate to avoid fatal loss (Ibid.). The story from past disasters is always kept in the memories of the people, and they maintain it as a lesson learned; thus, that can be a beneficial early warning.

The first examination of the SRG in the case of the 2018 Tsunami in Palu directly could identify the deep knowledge on the connection between local cultural and geographical context (Syamsuddin and Putra 2023). Several locations in Palu and its surroundings have been named depending on their geographical characteristics. For instance, 'Sigi, Biromaru,' means swamp, however, the development of the region that occurred shows that there is a neglect of the meaning stored in the values of local wisdom. The locations that are severely affected are those that have these names. However, the development of the region that occurred shows that there is a neglect of the messages within the values of local wisdom. The locations that are severely affected are those with the names of toponyms. The study conducted by Surtiari et al (2019) shows that people and policymakers are aware of the toponyms, but urbanization is growing without any concern for local wisdom.

3.4.2.2 Local Resilient Building Based on Traditional Architecture

Similarly, traditional local knowledge in Central Sulawesi is also found in the architecture of buildings that are resilient to earthquakes. Generated from Kaili's culture, the traditional buildings experienced less damage during the devastating earthquake followed by the tsunami in 2018. In contrast, the modern buildings were destroyed. Although in the end it was realized that there was an earthquake-resistant building model, during the reconstruction phase, earthquake-resistant buildings did not follow local wisdom but followed modern building models.

These are basically made of bamboo and wood with roofs thatched with alang-alang reeds. The single-story, stilted structure reduces the impact of earthquake vibrations because its potential for shaking is minimal structures (Saptaningtyas et al. 2023; Susanthi, Meisandi). For instance, flexible building design-concept has been considered in earthquake-resistant construction for post-disaster housing development in Lombok back in 2018 to create a building which would be more stable against earthquakes. While traditional construction is preferred, modern technology is used in reinforcing structural construction. To enhance resiliency against earthquakes, the study indicates the use of disaster-resistant construction approaches that incorporate conventional approaches with modern technology approaches. This applies to post-disaster housing development such as Lombok.

3.4.2.3 Warning System Through the Intergenerational Storytelling

Local wisdom in Simeulue Island, Aceh, known as *Smong*, has saved most of the island community from the giant tsunami in 2004. *Smong* is a natural warning system for the Simeulue people. The main message is that if a strong earthquake occurs followed by receded seawater, all members of the community have to run away to a higher and safer place. This wisdom is based on lessons learned by the community when they faced a mega earthquake and tsunami in 1907. This disaster killed most of the people on this island. The survivors then shared their experiences through storytelling (such as lullabies, traditional poems, and songs) that were inherited from generation to generation (Musfarayani 2009; Hidayati 2018).

The Simeulue people's experiences with extreme earthquakes and tsunami strongly motivate them to take action to lower the risks. Their experiences strongly influence memory, increase awareness and knowledge, and understand disaster impacts. These assist people's beliefs about whether and how they have to face disasters. They develop their capacities for engaging with others, through share-learning among peers and other members of the community. This engagement enables them to take relevant actions. Experience, belief, and preparedness are a close link to reducing disaster risk.

3.4.2.4 Mutual Assistance: Share Risk and Support Each Other

The Javanese in the District of Bantul, Yogyakarta, when faced with the devastating earthquake in 2006, utilized local wisdom for their survival and reduced the risk of further disaster impacts. It comprises community self-help *(tolong menolong*), mutual help/cooperation *(gotong royong),* and fundraising (*jimpitan*). The spirit of *gotong royong* made them strong and caring among neighbors. The fervor within a community and values of harmony and togetherness are the key elements in facilitating the effectiveness of the recovery efforts during the emergency and reconstruction phases.

This paper reveals that cooperation (*gotong royong)* and helpfulness (*tolong menolong*) among members of the community still exist and are practiced in most locations. The fervor of helpfulness (*tolong menolong*) is for personal or family gain, with the main characteristic being mutual assistance. A community member, for example, helps his neighbor in clearing his farmland and in return receives support from the neighbor. This local wisdom is crucial social capital for the community members to reduce disaster risk by increasing their preparedness. Meanwhile, the fervor of cooperation (*gotong royong*) is the joint bearing of burdens among the community members with volunteerism characteristics. The main objective is to attain community goals. For example, villagers work together in building meeting places or cleaning water channels to reduce the risk of flooding. They could apply *gotong royong* to set up, for instance, evacuation places and routes and to practice evacuation skills and drills.

3.4.2.5 Religious Leaders as Main References for an Early Warning System

In terms of disaster preparedness through an early warning system, the success of implementation is influenced by several factors both the technology and the communication system. Setiadi (2014) reveals that people at risk in Padang City follow the instructions and guidance from religious leaders rather than the warnings from the sirens. It is well acknowledged that Indonesian people have a strong transcendental belief when they have to deal with risks to natural hazards. For instance, they believe God will save them if they have good conduct. Moreover, some communities prefer to accept early warnings from the religious leader rather than a system from the government (Setiadi 2014).

3.5 Discussion

We use the approach of SRG to identify to what extent risk culture can be identified and integrated into the policy and program. Based on four major devastating disasters in Indonesia as explained above, it is obvious that Indonesian culture has various mitigation strategies to respond to natural hazards. However, the challenges are found when we examine the opportunity to be involved in the current risk governance of Indonesia, particularly in the stages of the fourth and fifth. The consideration of the local context in all processes of decision-making processes and consideration of cognitive aspects in the evaluation are still mismatched with the current government system in Indonesia. Moreover, the development of long-term scenarios will also need some validation to combine local context and scientific-based assessment.

Based on the six stages of systemic risk governance, it is clear that during the initial stage of social scanning, the community's local wisdom is easily identifiable and has already deeply ingrained itself into daily life. However, obstacles arise when it comes to long-term risk management planning. That is why stakeholders' capacity and institutional system continuously failed to favor the implementation of local wisdom into regulations and programs. For example, the consideration of toponyms conflicts with the short-term interests in economic growth. The return of infrastructure development to disaster-prone areas poses a new level of risk in the future. Similarly, the argument justifies the construction of houses without considering earthquake-resistant structures, citing the higher costs compared to modern house construction. Materials for wooden buildings and stilt house foundations are harder to obtain than modern building materials.

The long-term scenario can be learned from the local warning system in Simeleu namely Smong. Intergenerational risk communication works well because people are aware of the risks of living on the small island. The local people have common observations of nature such as animal behavior and sea level rise. The local government has tried to involve Smong in the disaster education material and dissemination. The integration of local warning systems into formal warning systems developed by national and local authorities is still challenging. Natural observation is currently facing an increased uncertainty that might result in less precise compared with the observation in the past. Therefore, accuracy and reliability of the local warning system still need to be strengthened by utilizing modern systems.

3.6 Conclusion

This paper aims to explore the consideration of risk culture to strengthen risk governance in Indonesia. This paper emphasizes the existing local wisdom to mitigate disasters, but it is still challenging to fully integrate into risk governance. We use the Systemic Risk Governance approach to answer our research questions. Deliberation of systemic risk is becoming more important to respond to the increased uncertainty and complexity.

Based on our analysis, the opportunities to integrate risk culture into risk governance to apply systemic risk governance are generated from enhancing risk awareness and improving the decision-making process. For instance, knowing toponyms collectively will influence risk perception to build better disaster resilience. The multi-stakeholders need to work collaboratively to apply the similar understanding they possess by prioritizing elements of fairness. The agreed-upon hazard-prone areas that are assigned as red zones should not be inhabited. All parties also must be adhered to by all parties, so that these areas are not used for residential centers or other public facilities.

Even though discussion on the role of risk culture has started among scientists and practitioners, policymakers still need to adjust to the government system. Measurement and evaluation that use cognitive indicators are still difficult to implement at the regional level. Cultural resistance will become another concern among the stakeholders that usually use non-culture indicators to govern risk.

We concluded that risk culture is important but yet still have no better understanding in terms of integration into Indonesian risk governance. Using steps to observe the applicability of SRG in Indonesia based on the observation of Triyanti et al (2023), it is clear that we have more opportunities to apply a systemic risk approach in Indonesia. Risk culture can be the facilitation regarding its nature as local context, inclusivity, and justice. Finally, we recommend continuously testing the SRG approach, particularly by promoting risk culture in the Indonesian case study so that the effectiveness of risk governance will become more possible.

References

Amri MR et al (2016) Risiko bencana indonesia. In: Jati R, Amri MR (eds). BNPB, Jakarta

Bakkour D et al (2015) The adaptive governance of natural disaster systems: insights from the 2010 mount Merapi eruption in Indonesia. Int J Dis Risk Reduct 13:167–188. https://doi.org/10.1016/j.ijdrr.2015.05.006

Birkmann J (2011) First- and second-order adaptation to natural hazards and extreme events in the context of climate change. Nat Hazards 58(2):811–840. https://doi.org/10.1007/s11069-011-9806-8

Birkmann J, Garschagen M, Setiadi N (2014) New challenges for adaptive urban governance in highly dynamic environments: revisiting planning systems and tools for

adaptive and strategic planning. Urban Clim 7:115–133. https://doi.org/10.1016/j.uclim.2014.01.006
Birkmann J, Setiadi N, Fiedler G (2015) A culture of resilience and preparedness: the 'last mile'case study of tsunami risk in Padang, Indonesia. In: Cultures and disasters. Routledge, London, pp 235–254
Bnpb (2013) Indonesia: national progress report on the implementation of the Hyogo framework for action (2011–2013) (October 2012), p 45
Bronen R, Chapin FS (2013) Adaptive governance and institutional strategies for climate-induced community relocations in Alaska. Proc Natl Acad Sci. https://doi.org/10.1073/pnas.1210508110
Burhany NR, Jiba A, Faradisa RB, Marwah PA (2022) 2. Architectural adaptation of vernacular stilt houses in Palu city. IOP Conf Ser. https://doi.org/10.1088/1755-1315/1075/1/012032
Cedergren A, Tehler H (2014) Studying risk governance using a design perspective. Saf Sci 68:89–98. https://doi.org/10.1016/j.ssci.2014.03.006
Cornia A, Dressel K, Pfeil P (2014) Risk cultures and dominant approaches towards disasters in seven European countries. J Risk Res. https://doi.org/10.1080/13669877.2014.961520
Cornia A, Dressel K, Pfeil P (2016) Risk cultures and dominant approaches towards disasters in seven European countries. J Risk Res 19(3):288–304. https://doi.org/10.1080/13669877.2014.961520
Djalante R, Holley C, Thomalla F (2011) Adaptive governance and managing resilience to natural hazards. Int J Dis Risk Sci 2(4):1–14. https://doi.org/10.1007/s13753-011-0015-6
Djalante R et al (2013) Pathways for adaptive and integrated disaster resilience. Nat Hazards 69(3):2105–2135. https://doi.org/10.1007/s11069-013-0797-5
Eriksen S et al (2011) When not every response to climate change is a good one: identifying principles for sustainable adaptation. Clim Develop 3(1):7–20. https://doi.org/10.3763/cdev.2010.0060
Mauelshagen C et al (2014) Effective risk governance for environmental policy making: a knowledge management perspective. Environ Sci Policy 41:23–32. https://doi.org/10.1016/j.envsci.2014.04.014
O'Brien KL, Leichenko RM (2000) Double exposure: assessing the impacts of climate change within the context of economic globalization. Global Environ Change 10(3):221–232. https://doi.org/10.1016/S0959-3780(00)00021-2
Pahl-Wostl C et al (2013) How multilevel societal learning processes facilitate transformative change: a comparative case study analysis on flood management. Ecol Soc. https://doi.org/10.5751/ES-05779-180458
Renn O (1998) The role of risk perception for risk management. Reliab Eng Syst Saf 59:49–62. https://doi.org/10.1016/S0951-8320(97)00119-1
Renn O (2008) Risk governance: coping with uncertainty in a complex world. Earthscan, London. https://doi.org/10.1037/0033-2909.I26.1.78
Renn O (2012) Risk governance in a complex world. Encyclop Appl Ethics. https://doi.org/10.1016/b978-0-12-373932-2.00009-0
Resilience Alliance (2010) Assessing resilience in social-ecological systems: workbook for practitioners. Version 2.0., p 54
Saleem M, Malik H, Salenda A, Winarta D, Muhamad R (2022) 6. Implementation of the souraja house façade on the composition of the government building façade in Palu city. IOP Conf Ser Earth Environ Sci. https://doi.org/10.1088/1755-1315/1075/1/012044
Saptaningtyas RS et al (2023) Earthquake-resistant wooden connection system in Sasak traditional buildings in Sade Village, Lombok, Indonesia. Civ Eng Archit 11:1890–1901
Schmit JT, Roth K (1990) Cost effectiveness of risk management practices. J Risk Insur 3:455–470
Setiadi NJ (2014) Assessing people's early warning response capability to inform urban planning interventions to reduce vulnerability to tsunamis case study of Padang City, Indonesia Neysa Jacqueline Setiadi. Rheinischen Friedrich-Wilhelms-Universität, London
Surtiari GAK et al (2019) Penanganan Pascabencana di Palu, Sigi, dan Donggala: Pemulihan Tempat Tinggal dan Penghidupan. Pusat Penelitian Kependudukan, Lembaga Ilmu Pengetahuan Indonesia (PPK-LIPI), Jakarta
Susanthi Y, Meisandy RR, Nisa A (2022) Earthquake mitigation based on local wisdom: the vernacular architecture concept of Dasan Beleq traditional house in North Lombok-Indonesia. GeoEco 8(1):48–61
Syamsuddin S, Putra RA (2023) 1. An ethnosemantic study of Palu Toponym and its association with natural hazards. JELTL J Engl Lang Teach Ling. https://doi.org/10.21462/jeltl.v8i1.1010
Tong F, Yang Y, Hongru Z, Zhenxing M (2023) 7. Risk culture as a blessing in tourism development: long-run effects of epidemic disasters. Ann Tour Res. https://doi.org/10.1016/j.annals.2023.103682
Tracy M, Marshall D (2020) 10. Risk perception and safety culture: tools for improving the implementation of disaster risk reduction strategies. Int J Dis Risk Reduct. https://doi.org/10.1016/J.IJDRR.2020.101557
Triyanti A et al (2023) Governing systemic and cascading disaster risk in Indonesia: where do we stand and future outlook. Dis Prevent Manag Int J 32(1):27–48. https://doi.org/10.1108/DPM-07-2022-0156
UNISDR (2009) UNISDR terminology on disaster risk reduction. International Stratergy for Disaster Reduction (ISDR), pp 1–30
UNISDR (2015) Sendai framework for disaster risk reduction 2015–2030 1', third world conference on disaster risk reduction, Sendai, Japan, 14–18 March 2015 (Preprint). Available at: A/CONF.224/CRP.1
Valentina L, Damienne P, Nathalie V, Cyrille B, Edwige D-P, Oscar N, Rodolphe C, Irmand M, Aziz-Alaoui MA, Abdel H, Boudoukha A, Schleyer-Lindenmann A, Berred SH, Emmanuel T (2023) 8. Modeling and analysis of the impact of risk culture on human behavior

during a catastrophic event. Sustainability. https://doi.org/10.3390/su151411063
van Asselt M, Renn O (2011) Risk governance. J Risk Res 14(4):431–449. https://doi.org/10.1080/13669877.2011.553730
Vinod K (2024) 2. Traditional and modern disaster resistant house construction techniques. J Prog Eng Phys Sci. https://doi.org/10.56397/jpeps.2024.03.04
Wachinger G et al (2013) The risk perception paradox-implications for governance and communication of natural hazards. Risk Anal. https://doi.org/10.1111/j.1539-6924.2012.01942.x
Zwick M, Renn O (2002) Perception and evaluation of risks. Available at: https://doi.org/10.1145/1900520.1900533

4 Unraveling Mental Health Issues and Psychosocial Support Post-disaster in Central Sulawesi

Abdul Fikri Angga Reksa and I Putu Ardika Yana

Abstract

This chapter examines mental health issues and psychosocial support in the post-disaster context of Central Sulawesi. Despite 5 years since the series of cascading disasters occurred, various challenges related to rehabilitation and reconstruction persist. One of the emerging issues that have not been widely discussed is mental health and post-disaster trauma among survivors. Before the disaster on 28 September 2018, mental health concerns were not extensively addressed by the Central Sulawesi community, largely due to societal stigmas. However, the recovery of mental health among survivors is a crucial aspect of post-disaster rehabilitation, alongside physical infrastructure development and economic recovery. Moreover, the survivors must confront additional systemic risks, such as the COVID-19 pandemic. The complexity of the disasters profoundly affected the mental well-being of the population. Hence, this chapter attempts to understand the impact of the disaster on mental health outcomes including its challenges. It also investigates the psychosocial support aspect, focusing on the influence of social networks, social cohesion, and social trust on individual and collective well-being in the aftermath of the disaster. This paper employs a qualitative method with a case study approach to elaborate on the current mental health challenges and post-disaster trauma among survivors in Palu City and Sigi Regency. The key findings indicate an escalation of mental health disorders from mild to severe symptoms, such as suicide incidents in the temporary shelters. Various initiatives from the government, non-government organizations, and the community have been implemented to establish community-based psychosocial support. However, the shortage of clinical psychologists and the negative stigma surrounding mental health make the community reluctant to address mental health issues. The study underscores the importance of implementing psychosocial support programs, engaging the community, and integrating mental health services within the broader disaster response and recovery framework.

Keywords

Mental health · Disaster risks · Psychosocial support · Central Sulawesi

A. F. A. Reksa (✉)
Research Center for Area Studies, National Research and Innovation Agency (BRIN), Jakarta, Indonesia
e-mail: abdu061@brin.go.id

I. P. A. Yana
Lembaga Sejenakhening.com, Palu, Indonesia
e-mail: ardika.yana@gmail.com

D. Hidayati et al. (eds.), *Examining Disaster Risk Reduction in Indonesia*,
Sustainable Development Goals Series, https://doi.org/10.1007/978-981-96-4182-6_4

4.1 Introduction

A series of major disasters occurred on 28 September 2018, devastating the province of Central Sulawesi. The National Disaster Management Agency (BNPB) recorded that the catastrophic event claimed the lives of more than 4340 people, with thousands more reported missing. Numerous buildings were destroyed, totaling 68,451 housing units and other hard infrastructure. The estimated losses from the disasters were projected to exceed IDR 18 trillion. Despite the passage of 5 years, issues related to the post-disaster recovery process still need to be fully resolved. One of the most urgent issues is the presence of residents who have yet to secure permanent housing. The availability of permanent housing is crucial for ensuring a seamless transition toward the *build-back better and safer* phase (Kennedy et al. 2008).

As of October 2023, the Regional Disaster Management Agency (BPBD) of Palu City estimates that hundreds of households are still residing in temporary shelters. They are dispersed across the regions of Palu City, Sigi Regency, and Donggala Regency. The encouraging news is that a significant number of survivors are patiently awaiting the construction of permanent housing in Phase II, totaling 4053 units. The entire project is planned to be completed gradually between December 2023 and March 2024. However, regrettably, there are also survivors whose fate remains uncertain. They have not received assurance from the government regarding the availability of permanent housing due to issues related to population administration, the distance of permanent housing from survivors' livelihood sources, and the complexity of land acquisition processes. According to Nursamsir et al. (2022), the provision of permanent housing is intricately tied to political processes and negotiations at both the local and national levels, often hindering the acceleration of housing development. Consequently, survivors are compelled to occupy crowded and uninhabitable temporary shelters, giving rise to social, economic, cultural, political, and health-related challenges, particularly during the prolonged COVID-19 pandemic (Nur et al. 2021; Reksa 2020; Tjitrawati and Romadhona 2023).

The unresolved infrastructure issues during the disaster recovery phase have given rise to far more complex problems. One less-discussed issue is the mental health problems experienced by disaster survivors. Mental health issues, such as post-disaster trauma, often go unnoticed and are frequently overlooked. In reality, post-disaster mental health problems occur with varying intensities. Numerous case studies from previous disasters highlight the importance of mental health aspects. For instance, research by Jenkins and Meltzer (2012) indicates that survivors of the 2004 Indian Ocean tsunami experienced various mental health issues, including post-traumatic stress disorder (PTSD), anxiety, and depression. On top of that, the study suggests that those who were forced to relocate from their original residences were at a greater risk of experiencing depression, particularly among women and children. Drawing from other cases, survivors of the atomic bombings in Hiroshima and Nagasaki in 1945 continued to experience physical consequences and psychological pressures even after 50 years had passed since the tragedy (Kokai et al. 2004). This illustrates that the post-disaster mental health recovery process is prolonged and ongoing.

Furthermore, how is the mental health condition of survivors in Central Sulawesi? Similar to disaster survivors in other areas, mental health issues have emerged within the Central Sulawesi community. Herdiana and Lakoro (2022), focusing on the psychosocial aspects of adolescents, revealed three psychological problems faced by adolescent survivors: trauma, anxiety, and grief. However, the study indicates that adolescents tend to address their psychological issues more promptly due to support from family and school. Nevertheless, the study does not elaborate on how parenting practices contribute to family resilience. On the other hand, a quantitative study involving disaster survivor subjects, who are also students at Tadulako

University, showed that 51.75% of survivors experience symptoms of anxiety (Rasido and Patodo 2020). There is no significant difference in terms of gender, indicating that both females and males are equally susceptible to high levels of anxiety. However, there is still limited research addressing how psychosocial aspects can be sustained in the context of disaster survivors.

In essence, the World Health Organization (WHO) has provided technical guidelines regarding mental health and psychosocial support (MHPS) in disaster and emergency conditions. In its report, the WHO emphasizes the importance of engaging communities to facilitate self-reliance in mental health, as well as identifying and protecting vulnerable groups (WHO 2011). On a national level, the Government of Indonesia has addressed MHPS issues outlined in Law Number 24 of 2007 concerning Disaster Management. Article 53 stipulates that during disaster management in emergency response situations, the community has the right to basic needs, including health services and psychosocial services. This further underscores the urgency of MHPS issues, although the implementation process is still hindered by various challenges, such as the lack of integration between mental health services and general health services, a shortage of medical personnel, and social stigma associated with mental health itself (Guilaran and Nguyen 2020).

Based on the above explanation, this article aims to discuss the handling of post-disaster trauma and mental health problems among disaster survivors in Central Sulawesi, especially in Palu City and Sigi Regency. The article focuses on discussing disaster survivors who have lived in temporary shelters and those who are still residing in temporary shelters. There are two overarching questions to be explored in this study. First, how does a series of disasters affect the mental condition of survivors living in temporary shelters? Second, how do disaster survivors build resilience in the post-disaster phase? The discussion on mental health aspects in the post-disaster phase in this section contributes to bridging a literature gap, which currently tends to concentrate more on infrastructure recovery and livelihood sources.

4.2 Methods

This research focuses on two disaster-affected locations on 28 September 2018 namely Palu City and Sigi Regency. To address the research questions, the study employs a qualitative method with a case study approach. Creswell (2014) defines the case study approach as an in-depth exploration of a program, event, activity, process, and one or more individuals within a specific time frame. The researcher collects detailed information using various data collection procedures over a sustained period. Data collection and direct observations were conducted longitudinally during the pandemic (21 September–2 October 2021) and post-pandemic (1–6 November 2023). The selection of these two periods aims to understand survivors' perceptions in facing various systemic and cascading risks. For data collection, the researcher utilized in-depth interviews and semi-structured interviews with disaster survivors who have lived or are still residing in temporary shelters. Additionally, direct observations related to the conditions experienced by survivors were conducted. In this article, the researcher collaborated with sejenakhening.com, an active community providing mental health services to survivors. They furthermore, played a role in critically analyzing the collected data in conjunction with other field findings. This paper also utilizes secondary data obtained from, literature review, government reports such as the Basic Health Research (Riskedas), policy documents, technical reports on NGO activities, and international frameworks related to mental health and psychosocial support.

4.3 Results

4.3.1 Mental Health Challenges Post-disasters

Not even 2 years have passed since the tragedy of the tsunami disaster struck residents living in the Talise Beach area, precisely on Cumi-Cumi Street. Their homes were destroyed and leveled to the ground. Many of them also lost family members. Around February 2020, residents were shocked to hear about a new threat in the form of COVID-19 in Indonesia. Initially, they were not daunted, never thinking that the virus would reach Palu City. However, their prediction proved wrong. On 26 March 2020, the Palu City Health Office confirmed the presence of a patient infected with COVID-19. Since the emergence of the first case, the lives of the shelter residents in Lere Village have never been the same. They had to adhere to the health protocols set by the government. There was unrest stirring in the hearts of the shelter residents, but there was not much they could do. They could only resign themselves to waiting for assistance to arrive.

For disaster survivors, government recommendations seem impractical. How so? Engaging in social distancing is unfeasible for them as they live in cramped chambers measuring 3 × 4 m made of gypsum board. Typically, one chamber is inhabited by more than one family head with family members ranging from two to eight persons, or even more. There is virtually no space between one chamber and another. They must share bathrooms with limited access to clean water. This implies that residents in temporary shelters face difficulties adhering to government recommendations for more frequent handwashing. This adds to the concerns of the community, given the reported casualties due to COVID-19 in Palu City. The misery of the survivors is further compounded as the aid received by the community diminishes over time. In the first 2 years post-disaster, survivors received various forms of social assistance from the government, local and international NGOs, and private entities. However, entering the third year, assistance continues to decline as donor funding has been depleted. Amid this uncertain situation, survivors can only hope to promptly obtain the promised permanent housing. Additionally, they seek assurance regarding economic sustainability. As identified, some survivors are still engaged in uncertain jobs due to increasing difficulty in finding employment commensurate with their competencies.

The consecutive disasters have significantly impacted the mental health of survivors in temporary shelters. One tangible manifestation is evident in several suicide cases within these shelters. Suicides are believed to be triggered primarily by economic pressures and depression. The non-profit organization Sulteng Bergerak has documented the first suicide case in a temporary shelter, which occurred in Petobo Shelter on 26 March 2019. A 24-year-old man was found hanging in a school building. According to the family, the man became despondent and withdrawn after his grandmother fell victim to liquefaction, and her body could not be found until now. Another case transpired in the Koni Shelter in Mantikulore, where a mother around 30 years old was found dead after ingesting poison in November 2020. Similar motives were discovered in the Lere Shelter, where a 30-year-old mother decided to end her life by hanging, despite being 7 months pregnant with two children. Economic factors, coupled with family issues, are strongly suspected to be the reasons behind her decision to take her own life.

The emergence of suicide cases in temporary shelters is one of the implications of the delayed post-disaster response in Central Sulawesi. Adequate and appropriate healthcare facilities should ideally be a priority in the disaster recovery phase. However, the majority of shelter residents lack sufficient access to both physical and mental health services. The ongoing inadequacy in guaranteeing healthcare facilities has led to an increasing sense of distress within the community. One informant admitted to still experiencing profound sadness, even though the tsunami tragedy occurred over 3 years ago. She lost her youngest child, who was swept away by the tsunami, and the body has not been found. To this day, she remains traumatized by the sight of the sea and feels anxious during heavy rain accompanied by strong winds. Furthermore, her fear intensifies at the prospect of not receiving a share of permanent housing in future.

> We have nothing left. We cannot return to Talise because it has been designated as a red zone. Meanwhile, the promised permanent housing has no clear timeline for realization. That's what makes us frustrated. I feel numb. (Ms. R, personal communication, 30 September 2021).

One health worker at the Lere Community Health Center (Puskesmas) mentioned that post-disaster mental health issues in Palu are mostly characterized by anxiety disorders, depression, and post-traumatic stress disorder (Personal communication, 8 November 2023). However, there is no precise data on the number of residents experiencing mental health disorders in the post-disaster period. This is due to the ongoing lack of public understanding regarding mental health. There is a prevailing stigma in the community associating mental health disorders with Persons with Mental Disorders (ODGJ), causing people to feel ashamed when seeking psychological services. The community is not fully aware that issues such as insomnia, excessive anxiety, and trauma fall under mental health disorders that require certain treatment. As a result, individuals with suspected mental health issues tend to keep their struggles to themselves and may even withdraw from social interactions. In situations of stress and a lack of assistance from anyone, suicidal decisions can occur. Regardless, programs related to mental health recovery have not yet become a priority in Central Sulawesi.

Similar findings were also disclosed by the institution sejenakhening.com, which actively provides psychosocial support in several disaster-affected areas. The public perception of mental health remains very limited. Findings in Sigi Regency, specifically in the villages of Sidera, Soulowe, Karawana, and Potoya, indicate that the majority of community knowledge about mental health is still confined to Persons with Mental Disorders (ODGJ). There are even those among them who have never heard information related to mental health services at the nearest health facilities. Consequently, when disasters occur, they are not aware of the mental health disruptions they experience. They perceive stress, sadness, and fear as normal and expect to fade away on their own. Moreover, rural communities tend to be unaccustomed to discussing matters considered private. People find it easier to express physical complaints than the emotional or psychological feelings they experience. As stated by one informant who admitted feeling uncomfortable sharing his feelings of sadness with strangers.

> I find it neither easy nor necessary to express how shattered my feelings are. In the first two months, I cried every day due to continuous sadness, and it persisted for a year, two years, and even now. [...] I prefer to surrender to Allah rather than to humans. (Ms. S, personal communication, 26 September 2021).

Among the mental health issues that emerge post-disaster, drug abuse has become increasingly prevalent. The National Narcotics Agency (BNN) of Palu City reported a 0.3% increase in the prevalence of drug users between 2021 and 2023. Although directly associating this rise in drug abuse with the disaster is not straightforward, the BNN underscores the presence of drug traffickers residing in temporary shelter tents. The implications of unstable mental health are also evident in the escalating cases of gender-based and sexual violence during disasters. The Human Rights Violations Victim Solidarity (SKP HAM) report recorded 125 cases of violence against women and child survivors of disasters and 90 cases of domestic violence (SKP HAM 2020). This warrants attention from the government and all stakeholders to protect survivors of gender-based and sexual violence, ensuring thorough investigations and justice for the survivors.

The mental health challenges faced by post-disaster communities are highly complex and closely related to social, economic, and cultural aspects. Even after securing permanent housing, there is no guarantee of psychological freedom from mental health disruptions. For example, survivors of liquefaction in Petobo, South Palu, now residing in Buddha Tzu Chi Housing in Tondo, Mantikulore, Kota Palu, still grapple with panic attacks triggered by loud noises or heavy rainfall. Some informants admit their inability to recount the painful tragedies they experienced. Ms. T, a retired private sector employee, consistently blames herself for not being able to save her 7-year-old grandson from the fatal consequences of liquefaction, even 5 years after the incident. However, she does not perceive this trauma as a serious issue requiring specialized treatment. Nevertheless, she remains open to consulting clinical psychology services if offered.

> The intricacies of mental health services remain unfamiliar to me. Are they costly? If so, perhaps

> I am fine with the current situation. I just attempt to find peace within myself and accept it as God's will. (Ms. T, personal communication, 5 November 2023).

Despite the limited awareness among the populace, mental health facilities in Central Sulawesi are notably scarce. As an example, the community in Sigi had no direct access to clinical psychologist services at the nearest community health center prior to the disaster. The constraints of healthcare facilities providing clinical psychology should be noted during Central Sulawesi's disaster reconstruction and rehabilitation phase. As highlighted by sejenakhening.com, mental health services cannot be optimally delivered and accessed even under normal circumstances, let alone in crises such as the post-disaster period (Butar-Butar et al. 2020). Before the disaster, mental health care was primarily administered at RSUD Madani in Palu City. However, as previously explained, understanding clinical psychology service facilities is not widespread at the grassroots level. People are still reluctant to seek mental health assessments due to various societal stigmas. When a disaster occurs, the community receives more extensive interventions from multiple entities such as the government, NGOs, and collaborating universities providing psychological first aid (PFA). Stakeholders concerned about the psychological well-being of the community make efforts to engage with residents in temporary shelters. Some NGOs and universities also conduct PFA training for health cadres, city government employees, and the community itself. In future, more people will provide psychosocial support independently, particularly to foster a sense of solidarity and a shared fate.

4.3.2 Invisible Battle of Mental Health

The issue of mental health is often perceived as invisible, although its implications are tangible. The fact that survivors decide to end their lives is adequate evidence that mental health problems require specialized attention. By prioritizing mental health issues, post-disaster recovery efforts can be more holistic and sustainable. The goal is to ensure survivors recover physically, mentally, and emotionally. In the context of the post-disaster situation in Palu and Sigi, the community appears resigned in facing various challenges such as PTSD, anxiety disorders, and panic attacks. However, this does not mean that the community is inactive in the struggle against post-disaster mental health issues. Survivors leverage social capital such as social cohesion, trust, and social networks to build resilience. Moreover, after the COVID-19 pandemic, the community is much more aware of mental health issues.

One example of the functioning of social networks and social cohesion in addressing mental health issues is evident in Talise's community of disaster survivors. During every congregational prayer and recitation session, they try to gather and share their feelings and grievances. Although they may not find practical solutions to their concerns, such gatherings serve as a means to release their stress. Undeniably, self-care with a spiritual or religious approach has become a part of the community's daily life. For the people in Talise, spiritualism not only serves a curative purpose but also plays a preventive role in dealing with more severe mental disturbances. According to one spiritual leader in Talise, by constantly remembering God, survivors can continue their lives with a sense of tranquility (Mr. R, Personal Communication, 3 November 2023).

Post-disaster mental health recovery aims to strengthen community resilience by emphasizing the importance of individual psychological resilience. The sejenakhening.com institution in Sigi Regency innovates to provide mental health services by professional healthcare workers in community health centers (Puskesmas) and trained personnel in villages. They deploy two clinical psychologists to Biromaru and Dolo Village and eight trained mental health village cadres in Sidera, Potoya, Karawan, and Soluowe. Since 2016, sejenakhening.com has been providing services for mental health problem recovery, transforming thought/perspective patterns, and altering maladaptive behavior through experiential learning interventions, group counseling, and individual counseling (Butar-Butar et al. 2020). To reduce the risk of future disaster-related mental health issues, efforts are made to establish disaster risk reduction documents based on mental health to serve as guidelines for villages, health departments, and public health centers responding to disaster situations.

The experience of one mental health cadre in Sidera Village, Ms. E, reveals that it is not easy to raise awareness among the community regarding mental health. Introducing counseling services to the community is a challenging task for the mental health cadre. It requires a personal approach to encourage the community to utilize the available counseling services. Erni's outreach activities are primarily focused on integrated health services (Posyandu), religious study communities, and village head office meetings. Initially, the community showed limited responsiveness to the counseling services, but over time, they began to make use of them. Predominantly, mental health cadres not only serve as agents introducing the concept of mental health issues but also act as "story companions" for those in need (Ms. E, personal communication, 25 November 2021). Nevertheless, in the event of a severe condition, the client will be directed to a healthcare professional at either the health center or hospital.

Apart from spiritual leaders and mental health professionals, the role of local leaders or local champions is crucial in the context of combating mental health issues. In the case of permanent housing in Mamboro, the role of the neighborhood leader (RT) is quite dominant, not only in ensuring the economic well-being of residents but also in maintaining their mental well-being. The RT fulfills the duty of being always ready to receive complaints and criticism from residents in the independently built housing (Huntap Mandiri). It is worth mentioning that former coastal residents of Mamboro initiated the construction of permanent housing in 2019 with the assistance of the Yayasan Arsitek Komunitas or Arkom Indonesia. Since the first construction of permanent housing, Arkom Indonesia has not only served as a consultant but also ensured that 39 households could regain their capabilities (physically, mentally, and economically). The community is now able to resume normal living with a highly robust support system at the neighborhood level.

The concept of constructing permanent housing was initiated with consideration for the aspirations of each resident and implemented collectively. The strong collective culture of mutual assistance (*Gotong Royong*) has helped the residents of Mamboro Housing feel more content with their lives. Most residents, who are primarily fishermen and engaged in small-scale fish-drying businesses, have been able to maintain their livelihoods due to the selection of a location not far from the coastal area. As a result, Mamboro Housing won the World Habitat Award Bronze Winner in 2021 as a collective housing structure for improved habitat recognized by the United Nations Human Settlements Programme (UN-Habitat). The housing project serves as a significant example of the importance of involving the community in post-disaster reconstruction and rehabilitation processes as mentioned by the Head of RT:

> We are not only concerned about the economic sustainability of the 39 households residing here but also ensuring their rights as citizens are fulfilled. This includes maintaining harmony, coherence, and community well-being. (Head of RT, Personal communication, 5 December 2023).

4.4 Discussion: Mainstreaming Mental Health and Psychosocial Support

Enhancing individual psychological resilience has become a crucial component of the community resilience approach to disaster management, specifically in Central Sulawesi. According to Basic Health Research (Riskesdas) in 2013 and 2018, Central Sulawesi Province has a track record as the province with the highest number of emotional disorders in Indonesia. Among the 34 provinces, Central Sulawesi holds the top position for the prevalence of the population with emotional disorders and the prevalence of those suffering from depression (Kementerian Kesehatan RI 2018). This should serve as a notable record for the government to pay closer attention to the mental health of the population. Moreover, based on previous research, severe mental health disorders have emerged in post-disaster regions (Sattler et al. 2018). Such incidents have indeed occurred in Central Sulawesi, as evidenced by the discovery of disaster survivors deciding to take their own lives. Although the issue of mental health has not resonated significantly at the grassroots level.

There are two main findings regarding the lack of awareness among the communities in Palu and Sigi about mental health issues, even though they have experienced them. First, the inadequate facilities and a significantly low number of mental health professionals or clinical psychologists contribute to this lack of awareness. Second, the community does not perceive mental health issues as crucial as physical health problems or other more visible issues such as economic challenges. Despite numerous efforts to mainstream post-disaster mental health issues and psychosocial support, most of them are still project-based, raising questions about its sustainability. However, the findings of this study have shown that social capital plays a crucial role in the disaster recovery process. To maintain community well-being, the established social networks have proven to assist the community in quickly recovering from the mental health disruptions they experienced. This proposition aligns with the research findings of Haslam et al. (2016), demonstrating that social identity interventions that build and strengthen social group membership enhance mental health. Albeit other studies caution against claiming social capital intervention as a preventive measure in cases of mental disorders due to a lack of evidence (Flores et al. 2018).

The disaster recovery framework, considering community-based mental health aspects, needs to be addressed following the findings of Trinidad and De-Castro (2020). Until now, regulations regarding post-disaster mental recovery refer to Law Number 24 of 2007. Besides, the Regulation of the Head of BNPB Number 11 of 2008 concerning Guidelines for Post-Disaster Rehabilitation and Reconstruction, also mentions the aspect of psychological and social recovery. The goal of psychological recovery is for the community to be able to perform social functions as before the disaster occurred, and to be prevented from experiencing psychological impacts, further leading to mental health disorders. However, all regulations do not entail specific technical guidelines on how mental health and psychosocial support should be done for disaster survivors. This has implications for post-disaster management that is less structured, especially in addressing mental health issues. The presence of both local and international NGOs, academics, and mental health practitioners is crucial in formulating tactical steps needed to prevent further deterioration.

To comprehend the systemic exposure of Palu and Sigi communities to risks, efforts to promote mental health and psychosocial support are crucial. The findings of this study indicate that traditional agencies, such as religious, spiritual, administrative, and cultural leaders, remain important in mental health problems. Therefore, to address the mental health issue more comprehensively, the involvement of community leaders is necessary. Additionally, training local cadres in every village or urban area to become mental health counselors is highly appropriate in the communal context of Central Sulawesi. The presence of mental health counselors serves not only during post-disaster periods but also supports clinical psychologists, whose numbers are still limited in Central Sulawesi. Previous research has revealed the effectiveness of the presence of mental health cadres in dismantling barriers among communities that have long been hindered by stigma regarding mental health (Windarwati et al. 2023). Therefore, specific guidelines on mental health must be tailored to the local context and its implementation.

Nevertheless, it is interesting to notice the post-pandemic conditions, which suggest an increased awareness of mental health issues, especially among the Gen Z population including in Palu and Sigi. Thanks to highly active social media influencers who advocate for mental health issues on Instagram, YouTube, and TikTok. One informant from the Gen Z demographic conveyed an understanding that challenges related to concentration during study and feelings of anxiety are classified as mild mental health disorders. This awareness was acquired through the information available on YouTube. Furthermore, they gained insights into self-help techniques, such as the Butterfly Hug method, designed to alleviate excessive anxiety or intense emotional outbursts (Ms. D, personal communication, 5 December 2023). Therefore, new media platforms, such as social media, can be utilized to enhance community awareness, which has been proven to be effective and efficient for mainstreaming mental health problems (Latha et al. 2020). Moreover, numerous studies have suggested that employing an artistic approach can impact the psychological recovery process for survivors, especially for children and adolescents (Orr 2007; Mohr 2014). However, adjustments should be made to incorporate local cultural art that is in harmony with the circumstances of Central Sulawesi.

4.5 Conclusion

The series of disasters that struck the Central Sulawesi region on 28 September 2018, still leaves deep wounds and griefs for the survivors. The ongoing challenges in the disaster rehabilitation and reconstruction phases have given rise to several issues, including mental health problems, which have not been addressed adequately. In reality, various efforts have been undertaken by the government, local and international NGOs, academics, and mental health practitioners to provide counseling services. However, these efforts are hindered by the limited availability of clinical psychologists in health centers and hospitals in Palu and Sigi. Additionally, various negative stigmas associated with mental disorders (ODGJ) discourage the community from seeking medical attention for their mental health conditions. The community tends to prioritize physical health over mental health, whether consciously or unconsciously they have been experiencing mental disorders. To build resilience and mental recovery, a structured framework and technical guidelines integrated with the local community or mental health-based disaster recovery are necessary. This is pivotal considering that Central Sulawesi holds the highest position as the province with the highest anxiety disorders in Indonesia based on the Basic Health Research (Riskesdas) in 2013 and 2018. The findings of this study indicate that the communities in Palu and Sigi rely on various forms of social capital as part of psychosocial support. Hence, it is expected that the government will pay more attention to the potential and strength of the community already present in constructing a more comprehensive framework.

References

Butar-Butar AR, Winarsih A, Yana IPA (2020) Merajut Karsa: Catatan Pemulihan Kesehatan Jiwa di Empat Desa, Dua Kecamatan, Kabupaten Sigi, Sulawesi Tengah. Yogyakarta: PT Nasya Expanding Management

Creswell JW (2014) Research design: qualitative, quantitative, and mixed method approaches, 4th edn. SAGE Publications, London

Flores EC, Fuhr DC, Bayer AM, Lescano AG, Thorogood N, Simms V (2018) Mental health impact of social capital interventions: a systematic review. Soc Psychiatry Psychiatric Epidemiol 53:107–119. https://doi.org/10.1007/s00127-017-1469-7

Guilaran J, Nguyen HA (2020) Mental health and psychosocial support services for disaster responders in Southeast Asia. In: Resistance, resilience, and recovery from disasters: perspectives from Southeast Asia. Emerald Publishing Limited, Berlin, pp 117–142. https://doi.org/10.1108/S2040-726220200000021008

Haslam C, Cruwys T, Haslam SA, Dingle G, Chang MXL (2016) Groups 4 health: evidence that a social-identity intervention that builds and strengthens social group membership improves mental health. J Affect Disord 194:188–195. https://doi.org/10.1016/j.jad.2016.01.010

Herdiana I, Lakoro R (2022) Psychosocial issues following natural disaster in Palu Central Sulawesi: a case study on adolescents. J Educ Health Community Psychol 11(2):424–442

Jenkins R, Meltzer H (2012) The mental health impacts of disasters. Government Office of Science, UK

Kementerian Kesehatan RI (2018) Hasil Utama Riskesdas 2018. Kemenkes RI. https://kesmas.kemkes.go.id/assets/upload/dir_519d41d8cd98f00/files/Hasil-riskesdas-2018_1274.pdf

Kennedy J, Ashmore J, Babister E, Kelman I (2008) The meaning of 'build back better': evidence from post-tsunami Aceh and Sri Lanka. J Contingencies Crisis Manag 16(1):24–36. https://doi.org/10.1111/j.1468-5973.2008.00529.x

Kokai M, Fujii S, Shinfuku N, Edwards G (2004) Natural disaster and mental health in Asia. Psychiatry Clin Neurosci 58(2):110–116. https://doi.org/10.1111/j.1440-1819.2003.01203.x

Latha K, Meena KS, Pravitha MR, Dasgupta M, Chaturvedi SK (2020) Effective use of social media platforms for promotion of mental health awareness. J Educ Health Promotion. https://doi.org/10.4103/jehp.jehp_90_20

Mohr E (2014) Posttraumatic growth in youth survivors of a disaster: an arts-based research project. Art Ther 31(4):155–162. https://doi.org/10.1080/07421656.2015.963487

Nur R, Aulia U, Napirah MR, Vidiyanto V, Aryuni M, Hendra S, Rusydi M (2021) Effects of the covid-19 pandemic on the stress level of tsunami-affected communities living in temporary housing in Palu City-Indonesia. Open Access Macedonian J Med Sci 9(E):70–75. https://doi.org/10.3889/oamjms.2021.5584

Nursamsir N, Jamaluddin II, Iqbal M, Ismanto I (2022) The social and political aspects of permanent housing provision for earthquake Tsunami and Liquefaction Casualties in Palu City, Central Sulawesi. JKAP Jurnal Kebijakan Dan Administrasi Publik 26(2):85–101. https://doi.org/10.22146/jkap.54013

Orr PP (2007) Art therapy with children after a disaster: a content analysis. Arts Psychother 34(4):350–361. https://doi.org/10.1016/j.aip.2007.07.002

Rasido I, Patodo M (2020) Post disaster: earthquake, tsunami, liquefaction mental health prevalence of Tadulako University students. Enfermeria Clin 30:214–218. https://doi.org/10.1016/j.enfcli.2019.10.071

Reksa AFA (2020) Trapped Populations: Menangani Pandemi COVID-19 Untuk Penyintas Bencana di Kota Palu. Jurnal Kependudukan Indonesia. https://doi.org/10.14203/jki.v0i0.546

Sattler ND, Claramita M, Muskavage B (2018) Natural disasters in Indonesia: relationships among posttraumatic stress, resource loss, depression, social support, and posttraumatic growth. J Loss Trauma 23(5):351–365. https://doi.org/10.1080/15325024.2017.1415740

SKP HAM (Sept, 2020) Langkah Yang Dilakukan Dalam Memandirikan Perempuan Pasca 2 Tahun Bencana. Solidaritas Korban Pelanggaran HAM. https://monitoring.skp-ham.org/wp-content/uploads/2020/09/Memandirikan-Perempuan-2-Tahun-Pascabencana-DP3A-Sulteng.pdf

Tjitrawati AT, Romadhona MK (2023) Affliction in the post Palu disaster: state failure to implement human rights standard on disaster management. Cogent Soc Sci 9(1):2233255. https://doi.org/10.1080/23311886.2023.2233255

Trinidad AC, Protacio-De Castro E (2020) The institutionalization of mental health and psychosocial support in emergencies in Indonesia. Int J Dis Risk Reduct 51:101918. https://doi.org/10.1016/j.ijdrr.2020.101918

Windarwati HD, Susanti H, Brooks H, Wardani IY, Hasniah, Raya M, Sari H (2023) Lay community mental health workers (cadres) in Indonesian health services: a qualitative exploration of the views of people with mental health problems and their families. PLoS ONE 18(11):e0289943. https://doi.org/10.1371/journal.pone.0289943

World Health Organization (2011) Disaster risk management for health: mental health and psychosocial support. Global Platform, May 2011

5 Framing Disaster Preparedness: Media Perspectives

Temi Indriati Miranda

Abstract

Mass media serves four primary functions: informing, educating, entertaining, and persuading (Schramm in Mass media and national development—the role of information in the developing countries, Stanford University Press, Stanford, 1964). In the context of disasters, the media plays a critical role in delivering information, providing mitigation education, and offering persuasive content. However, limited resources and space sometimes prevent mass media outlets from fully realizing these functions. Often, market demands compel them to filter and prioritize the information they present to the public. In disaster coverage, the media may focus more on dramatic issues such as aid distribution controversies—where certain groups receive more assistance than others—rather than the disaster event itself. It appears that the entertainment value of such drama often overshadows the educational and informational aspects. On the other hand, the media are also expected to assist stakeholders by disseminating information about natural hazards, educating the public on risk mitigation strategies, and persuading audiences to remain vigilant against potential threats. To some extent, they have fulfilled this role, as people rely on the media to keep them informed about disasters across Indonesia. But does the media, particularly newspapers, contribute effectively to educating and persuading people about disaster risk reduction? This study aims to collect all online newspaper articles related to natural hazard-induced disasters—specifically tsunamis, earthquakes, and floods—published in Indonesia from 2005 to the present. Through qualitative content analysis, this chapter seeks to illustrate how the media communicates information provides mitigation guidance and persuades readers about disaster risk reduction and management.

Keywords

Disaster risk reduction · Content analysis · Mass media

5.1 Introduction

Indonesia is a vast archipelago consisting of thousands of islands. It is located in the Pacific Ring of Fire, home to hundreds of active volcanoes. Additionally, Indonesia sits at the convergence of three major tectonic plates: the Indo-Australian plate (in the south), the Pacific plate

T. I. Miranda (✉)
Research Center for Population—The National Research and Innovation Agency (BRIN), Jakarta, Indonesia
e-mail: temi002@brin.go.id

D. Hidayati et al. (eds.), *Examining Disaster Risk Reduction in Indonesia*,
Sustainable Development Goals Series, https://doi.org/10.1007/978-981-96-4182-6_5

(in the east), and the Eurasian plate (in the north). This geographical positioning makes Indonesia one of the country's most prone to disasters, including earthquakes, volcanic eruptions, and tsunamis (Fauzi and Marhamah 2020). Furthermore, Indonesia's tropical climate, characterized by wet and dry seasons, adds to its vulnerability. Extensive deforestation exacerbates these risks, increasing the likelihood of floods during the rainy season and droughts during the dry season.

Floods and droughts can be minimized by maintaining the natural balance and promoting reforestation. However, some natural hazard-induced disasters, such as earthquakes, volcanic eruptions, and tsunamis, cannot be avoided. Indonesia continues to struggle with these disasters, as evidenced by the relatively high number of deaths and affected population (Nasution 2018). Therefore, mitigation is crucial to addressing this problem. The involvement of multiple stakeholders is essential to support these efforts, with the mass media serving as one of the key components.

Moreover, Indonesia is urged to improve its disaster communication system, with one key component being the use of mass media (Khumairoh et al. 2021). Sharing knowledge and exercising social control are two primary functions of mass media (Donohue et al. 1973). In the context of disasters, mass media plays critical roles in transforming information, enforcing social control, and providing education. Ultimately, mass media is expected to raise public awareness about natural hazard-induced disasters (Khumairoh et al. 2021).

On the other hand, this is not an easy job, as hazards and disasters do not always occur in easily accessible locations. Moreover, in most affected areas, disasters cause broken roads and lost connectivity. Therefore, people rely on local journalists or local government staff for information. Despite these limitations and challenging conditions, journalists are still able to present information about the disasters, allowing people who live far away from the affected areas to receive the information.

When it comes to disaster preparedness and mitigation, the media has significant potential to contribute. However, media outlets often prioritize sensational stories over less dramatic but essential topics like disaster mitigation and preparedness (Bansal 2018). As part of a business industry, media organizations need captivating content to attract audiences and generate revenue. In Indonesia, this often translates into coverage of emotional victim stories, delays in aid delivery, and political aspects of international assistance, rather than focusing on preparedness efforts.

Despite this, the media plays a critical role in shaping public and decision-makers' priorities (McCombs and Valenzuela 2007). While sensational news attracts larger audiences, it carries a vital responsibility: promoting disaster preparedness. This paper seeks to analyze the role of mass media in disaster reporting in Indonesia. By examining the four functions of media—information, education, entertainment, and persuasion—this study explores which is most dominant through a content analysis of hundreds of online newspaper articles.

5.2 Literature Review

"The significant body of literature on journalism and disaster coalesces around three broad themes: disasters and their impact on news organisations and journalistic practice; the characteristics of reporting and news media coverage of disaster; and research that has explored media-audience dynamics at the different phases of disaster". (Matthews 2017: 3).

Mass media hold an important but often misunderstood role in disaster events (Sarma 2021). The media can play a critical role before, during, and after such incidents (Scanlon 2011). Moreover, media has an important role in disaster preparedness and management (Nahak 2021). As a rule, the journalistic reports contributed to the naturalization of disasters and the gap in public understanding of risks, by presenting an approach focused on relief actions—ignoring social issues, vulnerability, and population resilience—reducing the discourse of preparedness for future disasters (Antunes et al.

2022). So that, information provided in the newspapers could not have had a direct role in risk communication or, as a result, in changing people's behaviour (Tekeli-Yesil et al. 2018).

When a disaster happens, there are three dimensions of coverage by journalists: Impacts, factual, and dramatization; each dimension has some categories that have issues to be reported (Antunes et al. 2022). Dimension of impact covered three issues: socio-economy, health, and environmental. Factual includes description of event, voices, and sources, and search and rescue operations. Meanwhile, dramatization mentions personal drama, accountability/blame and risk, uncertainty, and fear (Antunes et al. 2022). This framework process could be implemented in any disaster journalists around the world. However, which categories more highlight is varied, depending on the interest of the mass media.

5.3 Methods

Content analysis is "Analysis of the manifest and latent content of a body of communicated material (as a book or film) through classification, tabulation, and evaluation of its key symbols and themes to ascertain its meaning and probable effect". (Krippendor 2004: XVII). Content analysis aims to organize and elicit meaning from the data collected and draw realistic conclusions from it (Bengtsson 2016). Meanwhile, content analysis is a highly flexible research method widely used in library and information science (LIS) studies with varying research goals and objectives (White and Marsh 2006). Beginning as a tool for mass communication studies, now many social sciences (sociology, anthropology, etc.) use content analysis as a method of study (White and Marsh 2006).

Frame of Process of Media report

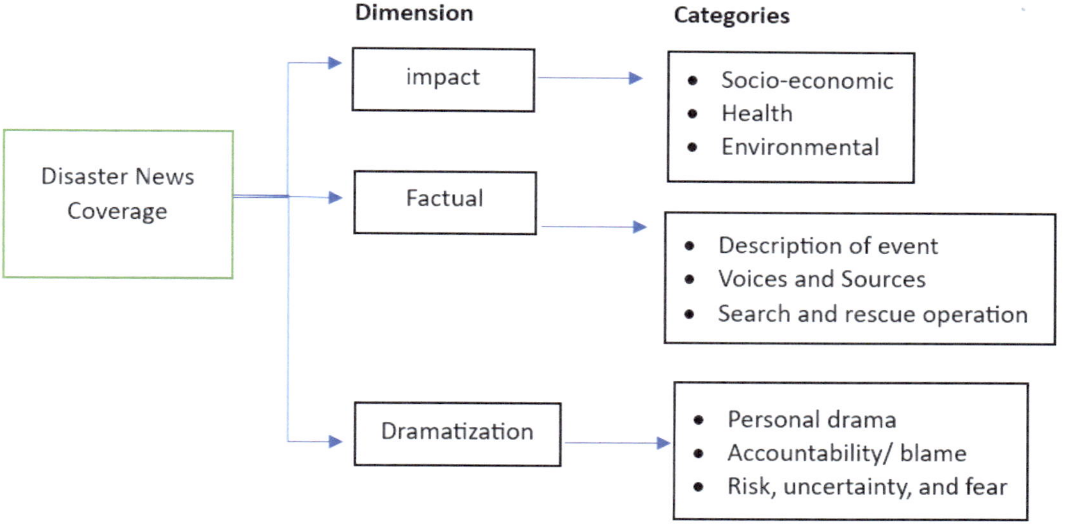

(adopt from Antunes, Pereira, Zêzere, and Oliveira, 2022).

The process of analyzing data for this writing is (1) listing online newspaper articles, the process of searching by using google as a search engine with keywords for any kind of disasters in Indonesia: Tsunami, earthquake, eruption, and flood and combining it by place, such as Tsunami Aceh, Tsunami Pangandaran, etc.; (2) categorizing the articles by kinds of disasters, places, and time; (3) reading all the articles in every category, find, and conclude which is the most highlighted function of the news (information, education, entertainment, and persuasive).

The process of data searching uses keywords: such as name of disaster, location, and time on the google search engine. After google displaces all the information, this research only chooses information from online newspapers, either local or national companies. All the information chosen is in Indonesian texts. The process of analyzing starts with categorizing by name of disasters such as "Tsunami Aceh", "earthquake Yogyakarta", etc. And every theme is categorized again based on the content of the news. The analysis ended after finding the flow of the information, getting the pattern of news, and finding which is the most frequent information that newspapers distributed to the readers. The newspapers online that are used are: Kompas.id, Kompas.com, Detik.com, Antara.com, republika.co.id, CNBCIndonesia.com, Sindonews.com, BBC.com, Liputan6.com, RCTIplus.com, nationaltempo.co, tribun.com, timesIndonesia.co.id, harianrakyataceh.com, Tirto.id, etc.

The information in the next part (result part) is the compilation of points of view that are paraphrased from many articles on online newspapers. Many articles distribute the same information and issues, if only sometimes, there are differences in numbers, such as a number of survivors or number of magnitude power of an earthquake. If this is the case, the writer took the most stated numbers in many articles' online newspapers or used only numbers of thousands or hundreds not mentioned in detail of units or tens digits. On the discussion part in every issue, the analysis based on results and literature is done.

5.4 Media's Perspective

5.4.1 Tsunami in Indonesia

Tsunami can travel at a speed of more than 900 km/h, especially those caused by earthquake at the bottom of the sea.[1] As we know, Tsunami term is from Japan's language: Tsu: Harbour and Nami: wave, this term was used for the first time in 1896 in East Coast Japan, when this coast was attacked by huge waves and earthquakes[2,3]. Not only in Japan or in Indonesia, Tsunami (either small or mega) went to many seas all over the earth[4,5]. Indonesia has experienced many Tsunamis. Noted, more than ten tsunamis in Indonesia caused more than hundreds of people killed[6,7,8,9] (data in Table 5.1) Moreover, media also mentioned that Indonesia has experienced Tsunamis since the year 416, and more than 200 Tsunamis have been detected in Indonesia ever since[10]. However, after Tsunami Aceh, Indonesia raises more concerns about this disaster, before the Tsunami Aceh, Indonesians only called Tsunami as "tidal wave".

Furthermore, this paper shows more about media framing of Tsunami in Indonesia from Tsunami Aceh to Tsunami Banten-Sunda Strait,

Table 5.1 Tsunami in Indonesia that caused many people killed

Place	Time	Victim
Aceh	2004	227.900
Lampung	1883	36.000
Flores	1992	2.500
Banda	1899	2.500
Banda	1674	2.200
Palu-Donggala	2018	4.845
Bali	1815	1.200
Pangandaran	2006	664
Mentawai	2010	456
Banten-Sunda Strait	2018	437
Banyuwangi	1994	250
Banggai	2000	264

(*Source* Data compiled from online Newspapers[11,12,13,14,15])

which are the natural disaster that happened after 2004: Tsunami Aceh (2004), Tsunami Pangandaran (2006), Tsunami Palu-Donggala (2018), and Tsunami Banten-Sunda Strait (2018).

5.4.1.1 Tsunami Aceh (2004)

It happened almost 19 years ago; it was on 26 December 2004[16]. It killed more than a hundred-thousand inhabitants in Aceh, North Sumatera, and its surrounding areas[17,18,19]. Tsunami Aceh is named as the biggest and the worst natural disaster in Indonesia[20,21]. The tsunami that happened in 2004 was not the first time in Aceh[22]; "Smong" is the local term (Simeleuh Island) named Tsunami[23]. For the last 5 years, almost every 26 December online media try to remind us about that day. There are many coverages related to the incident: rewrite the process of Tsunami that happened in that areas[24,25,26,27,28], try to remind us that Tsunami is always connected to the earthquake[29], the story of survivors[30,31] or many hoax issues related to Tsunami Aceh: Issue that Tsunami Aceh was a nuclear bomb[32].

Tsunami Aceh also had an impact on the conflict between Indonesia and GAM, a year after Tsunami Aceh, memorandum of understanding signed in Helsinki, 15 August 2005 and now, conflict between Indonesia and GAM reduces significantly; even, one GAM's member was back to be an Indonesia's citizen, and became the head of East Aceh Regency (2012–2022)[33,34]. There is always wisdom in every incident. Indonesia received many international aids to help Aceh, Nias, and North Sumatera at that time[35,36,37]. At least there are 18 countries and 17 Middle East countries sent their aids and medical staff to recover Aceh, North Sumatera, Nias islands, and the surroundings[38,39]. However, after several years passed many international aids changed allotment[40]. Many survivors could not get the aids regarding administration problem and land ownership[41]; on the other hand, many houses funded by international aids were abandoned due to the un-strategic location[42].

Along with time, the awareness of mitigation in Aceh has reduced, making it difficult for the volunteers to socializing the disaster mitigation[43,44]. The mitigation facilities and infrastructure are not used at their maximum levels, moreover, Aceh also lacks expert in disaster mitigation, even though hundreds of aids have been unstoppable to Aceh after Tsunami 2004[45,46].

5.4.1.2 Tsunami Pangandaran (2006)

On 17 July 2006, more than six hundred people killed[47], thousands of people evacuated to the safe places such as village halls, mosques, or schools[48,49]. First earthquake was at 15.11 WIB (PM) with 6.8 Richter scale, after that another earthquake: 5.5 Richter scale and 6.1 Richter scale; the coverage areas of Tsunami at that time were Pangandaran, Cilacap, Kebumen, Bantul, Pacitan, and Surabaya[50]. Similar to Tsunami Aceh, most articles on online newspapers are in memorial which means that articles published several years after the disasters, only few articles at the same year that the Tsunami happened.

In the memorial of Tsunami Pangandaran, many victims share their sad experiences regarding the Tsunami that happened on Monday, while days before (on the weekend), there was an international kite festival attended by many tourists[51]. Also, a hamlet is no longer inhabited after 35 residents were killed during the Tsunami; the residents became so traumatic due to that disaster that they decided not to stay in that area anymore, and now that area has become an aquarium park[52]. The village's myth (fatalism) about tsunami was also explored by media, and many witnesses confessed about their spiritual sights before that disaster happened[53,54].

Pangandaran also received aid from national and international[55,56,57] funds. However, 3 years later, there is news about the corruption of the Pangandaran Tsunami aids[58]. Regarding there are many corruption cases in disasters' aids, President SBY asked external auditors and aids registered to serve as control[59]. After Tsunami Pangandaran, both national and local governments did many things to recover the area: rebuilt some public facilities, relocated, and developed houses for the victims. They also socialized and educated inhabitants on evacuating if a tsunami happened[60]. Some mitigations are Pangandaran mangrove planning that developed brake-water[61].

5.4.1.3 Tsunami Palu-Donggala (2018)

More than 2000 people were killed, more than 10 thousand injured, 70 thousand people evacuated on 28 September 2018 at Tsunami and liquefaction Palu-Donggala[62,63]. This disaster was terrifying due to three disasters in 24 h (earthquake with 7.4 magnitude, tsunami, and liquefaction)[64,65]. So, it is categorized as a very rare event[66]. Before that, public in Indonesia was not aware of liquefaction; even though liquefaction in Palu was not the first time, this is the biggest one and all damage cost more than IDR 2 trillion[67]. The earthquake damaged the power grid causing many inhabitants to not be able to receive the early warning tsunami[68]. The drop of the electricity and the minimum of tools also become the obstacle of the evacuation; many volunteers only rely on their own energy to dig the debris and find the victims by their voices asking for help[69]. The survivors reminded each other about what they must do during the aftershocks: run to the open space, leave the house immediately, and stay away from trees and poles[70].

Indonesia also received more than IDR 220 billion international aids from more than 14 countries[71,72,73,74]. If only Indonesia's government was more careful and sorted out the aids better. They did not immediately welcome all aids and staff that came to Indonesia; in the name of national security, all kinds of aids coming to Indonesia needed to be registered[75]. There are many studies about mitigation of Tsunami at Palu-Sigi-Donggala (Damayanti et al. 2021; Pratomo and Rudiarto 2013; Sukino et al. 2019). Online newspapers support mitigation information[76], if only the ability of the locals to mitigate was not still low[77].

5.4.1.4 The Banten and Sunda Strait Tsunami (2018)

Located only two hours from Jakarta, (many affected areas in Sunda Strait are local tourist destination such as Anyer beach, Carita beach, Tanjung Lesung beach) and it happened on the weekend night (Saturday, 22 December 2018) and close to public holiday (Christmas), made this tragedy more dramatic. More than 400 people were killed and most of them are local tourists, the locals, hotel and resort workers, and entertainers (including celebrities and their families who performed at an event in that place)[78,79]. This disaster was triggered by the eruption of Krakatau Mountain located at Sunda Strait as a reason why tsunami warning system was failed, regarding the tsunami detector could only detect the tectonic earthquake not the eruption one[80,81,82,83].

Due to the damages and victims, the Indonesian government did not receive international aid for this tragedy. Most of aids is from the inter-local government, national government, and citizens and privates[84,85,86]. Early warning system and mitigation are very important in the beach area; therefore, stakeholders asked the entrepreneurs, or investors to develop hotels or resorts at the minimum of 500 M from the shoreline[87]. The condition of Anyer beach and Carita beach (area of Tsunami Banten 2018) as domestic tourist destination is bad now for hotels and restaurants' business, after two bad conditions: Tsunami 2018 and COVID-19. The occupation for hotel is only 20%[88,89]. So that the two tragedies have a long impact on the economic conditions for the locals who depend on their livelihood to the tourism industry.

5.4.1.5 Discussion About Tsunami in Indonesia Based on Mass Media Data

From the description about the four tsunamis in Indonesia above, this disaster brought huge damage to Indonesia. Due to numbers of victims killed and loss of properties. Media online (online newspapers) brought an important part of the information to the readers all over Indonesia (regarding the language they use). Mass media updated the numbers of victims, damages, story of the victims, and aids. This is along the line with what Antunes et al. (2022) mentioned about three dimensions of coverage by journalists: Impacts, factual, and dramatization.

On perspective of mitigations, the repetition of information on the newspapers about the earthquake comes before tsunami brings awareness to the public about the natural side on the tsunami. This is a success story, if only the role of mass media on mitigation requested more than that. The portion of mitigation in newspapers is still limited compare to impact, factual, and drama. Drama behind the disasters is also an issue to be distributed on online Newspapers. Tsunami Aceh brought a peace agreement between Indonesia's government versus GAM. The conflict had happened for more than 20 years. This is one of the samples of a happy ending drama that reported by mass media on disaster news. Other than that, there was also drama about personal stories of the victims, such as a celebrity who lost his wife during the Banten-Sunda Strait Tsunami, he described the process of survival from the disasters, and how he found his best friends and wife. It seems only a sad story, but from that experience, people learned how to survive in the middle of a disaster.

5.4.2 Earthquakes in Indonesia

Due to the typology and geography of Indonesia as mentioned earlier, Indonesia has experienced many earthquakes. BMKG (2023) (Meteorological, climatological, and geophysical agencies) reported for the last 2 months (November–December 2023), there were 30 cases of earthquakes in Indonesia. However, the magnitudes were around 5 and located at the oceans that far away from lands, therefore, it did not have impacts on the people (BMKG 2023). There are many kinds of earthquakes: that bring tsunami, or due to mountain eruptions, or movement of the earth's plates. In this session, earthquake-caused Tsunami were excluded since they were already discussed in the previous part. So, on mass media framing, this session describes any other news that the online newspapers highlight on earthquakes in Indonesia. Since the year 2000, there have been two big earthquakes that caused thousands of people killed: earthquake in Yogyakarta (2006) and earthquake in Padang (2009). Meanwhile, earthquake in Cianjur (2021) killed more than 600 inhabitants. So, this earthquake's session will discuss the three cases: Yogyakarta (2006), Padang (2009), and Cianjur (2021), and focus on what the highlights of online newspapers produce about the information to the public.

5.4.2.1 Yogyakarta (2006)

The earthquake happened at 5:53 AM, on 27 May 2006, with 5.9 Richter scale power[90,91]. The epicentrum of the earthquake was at the hills located ± 15 km to the east of Opak canal[92]. It only happened for 57 s, but killed more than 6000 people[93]. The most damage areas were Klaten, Central Java and Bantul, Yogyakarta[94]. According to BMKG, this earthquake was a warning that subducted zone was still there[95]. The expert from BMKG mentioned that earthquake was not dangerous, but the buildings that were effected of the earthquake were, so the buildings were the prone areas that needed to be resistant to the earthquake[96]. Locals were so shocked since they were focusing on mitigate of the Merapi eruption at that time[97]. Moreover, Indonesia's inhabitants still had trauma due to Tsunami Aceh (2004) that just happened 2 years prior. That traumatism caused the chaos condition regarding issues of tsunami that were coming after the earthquake in Yogyakarta[98].

Not only from the Indonesian governments, more than 34 countries distributed their aids to the victims of Yogyakarta earthquake; the aids were not only things such as foods, money, and medicine, but also international volunteers, doctors, and US marines went to Yogyakarta to help the life victims and the locals finding more victims in the ruins of the damaged areas[99,100,101]. Turkey's government also built 100 houses for the earthquake victims in Yogyakarta, while Indonesia's government built more than 1000 houses[102]. More than that, the international aids did not only help the victims by supporting them by giving them foods, medicines, or built infrastructure, and shelters, but the international aids (Germany, Dutch, England, Finland, Denmark, and World Bank) under an organization called Java Reconstruction Fund (JRF) help to revitalize the small, micro, and medium enterprises (MSME's), so that the victims did not only recovery both physically and mentally but also financially[103].

5.4.2.2 Padang (2009)

It was at 5:16 PM, 30 September 2009, an earthquake with a power 7.6 Ritcher scale damaged West Sumatera especially Padang city and its surrounding area, and the earthquake's power was felt throughout Sumatera, West Java, and Jakarta, and even to the neighbouring countries such as Singapore, Malaysia, and Brunei Darussalam[104,105]. It killed more than a thousand people, injured more than two thousand people, and destroyed more than 200.000 houses and buildings[106,107]. The most damaged areas are Padang pariaman, Padang, Agam, and Pariaman (they are all located in West Sumatera Province), and the earthquake's epicentrum was located ± 50 km of the northwest coast of Sumatera with ± 87 km deep from the surface[108].

After natural disaster events that cause many people to lose their live, there are always contemplation events in almost every year after; the purpose of that event is to mitigate locals and public on how to cope with that situation. In that event, the victims rememorized and talked about their feelings about the tragedy[109,110]. Due to this natural disaster more than a thousand people were killed and hundreds of people lost their homes and suffered, international aids came to this place. Indonesia received many aids from Australia, England, Qatar, Uni Arab Emirate, China, Japan, Germany, and Singapore. They did not only send the aids' logistic, but also teams to help find victims, and reduce the damages: such as doctors and paramedics, and SAR teams[111,112]. Indonesian government also spent IDR 2 trillion to rebuild 114 houses for the victims[113]. However, aids distribution was not smooth in these areas: damages and difficult areas and irresponsible people taking advantages from disasters by looting the aids that hinder the effectiveness of the aid's distribution[114].

5.4.2.3 Cianjur (2022)

According to BMKG, Cianjur and its surrounding are one of the earthquake's prone areas[115]. Earthquakes happen many times in Cianjur since the Dutch colony; it is a 20 years cycle, which means it happens in every 20 years and the impact of the earthquake was high based on the epicentrum of the earthquake that is relatively shallow ± 10 km to the land[116]. However, the tragedy in 21 November 2022 at 1:21 PM was a horrible disaster because 600 people were killed, more than 2000 people were injured, and thousands of buildings were destroyed[117,118,119]. The earthquake with magnitude power of 5.6 Ritcher Scale was caused by the fault's movement of Cimandiri caesarean[120].

In this disaster, Indonesia's government did not receive international aids, so many local governments (regencies), political parties, and companies distributed their aids to the victims. Not many newspapers reported these issues because they already have their own website to publish it[121,122,123,124,125,126,127,128,129]. The Cianjur Regent mentioned that Cianjur received many aids from many stakeholders, if only the aids just went directly to the impact areas without any controlled or registered, it seemed the aids were not distributed fairly, regarding many victims did not receive aids properly, on the other hand, other victims got a lot of aids[130].

5.4.2.4 Discussion About Disaster Online Newspaper's Articles: Earthquakes in Indonesia

Mass media plays an important role in collecting and distributing data to the public. The reports are always about information on victims: how many people were killed and injured, who they are, and how they survived after the disaster;

detail information about the disaster's time, location, the epicentrum, the magnitude power, and causes; impact areas and damages. Furthermore, aid is always an issue for online newspapers. Especially international aids, countries that give aids, kinds of aids they gave, and how much the aids are always important issues. Mass media also thought that was important to report the process of distribution, control, and management of aids. Because aids are not a debt, and therefore fragile to corruption and pillage. On the other hand, image is also important for certain parties. In many cases, mass media becomes a political vehicle for some parties on distribution of aids. As the messengers of mitigation to the public, mass media is expected to be as creative as possible. Rememorizing a disaster is always a good moment for mass media to distribute the mitigation. Recalling sad stories from the victims is the prologue for the journalists to the mitigation information. The mitigation/education of the disaster awareness distributed as simple as possible and repeated many times to stay in public's minds.

5.4.3 Flood in Indonesia

Flood is something else, and it is always been an issue in Indonesia. Every year Indonesia deals with flood, not only in rainy days, but also in dry season. Moreover, flood is the most frequent disaster in Indonesia[131]. Flood never be a purely natural disaster, in many cases, it has human contribution[132,133]. In 2022, it is noted that there were 847 cases of flood in Indonesia, and it brought losses annually of IDR 22 trillion to the country for the last 10 years[134]. More than that, floods that often come cause numerous losses, some areas reported their losses such as: Jakarta and its surrounding lost IDR 10 trillion due to flood in 2020[135]; Central Java also reported that they lost IDR 16 billion due to flood in early 2023[136]; Aceh also reported the loss of more than IDR 10 billion due to a week flood causing crops failure in 24 villages[137].

At least there are 20 cities in Indonesia that are flood prone: Jakarta, Bandung, Surabaya, Medan, Padang, Pekanbaru, Jambi, Bandar lampung, Pontianak, Samarinda, Makassar, Ambon, Manado, Gorontalo, Kendari, Palembang, Jayapura, Sorong dan Palu[138], all five biggest cities in Indonesia are included as area of flood prone. There are at least 5 cases of big floods in Jakarta that: January 1996 (20 people died), 1 February 2007 (80 people died), January 2013 (20 people died), January 2014 (23 people died), and 1 January 2020 (16 people died)[139]. Meanwhile, it is noted that there were big flood and flash flood in Indonesia that resulted in losses and damages to the livelihood and even worse, killed inhabitants: Sintang, West Borneo (November 2021) the duration of the flood was almost 1 month, Manado (January 2022) 18 people killed due to the disaster, East Flores, NTT (4 April 2021) 69 people died, Tangse, Aceh (March 2011) 24 people died, Wasior, West Papua (October 2010) 150 People died, Mandailing Natal, North Sumatera (October 2018) 12 People died, Jakarta (February 2007) 80 People died, and Bukit Lawang, North Sulawesi (November 2003) 200 people killed due to the disaster[140,141,142]. Not only losses of life, but flood also cause many losses to livelihood. There are at least five of losses due to flood: (1) Economically loss (2) Limited clean water (3) Health problems (4) Paralyzing the activities[143,144].

There are many things that both central government and locals' governments do to cope with these problems; however, they can only reduce the impact of the flooding areas, but cannot make this disaster disappear. Several suggestions and tips are socialized to mitigate floods to the public[145,146]. Local governments also develop and fix infrastructure as the process of flood mitigation such as Semarang and DKI Jakarta, Tangerang and West Jakarta[147,148,149]. Not only infrastructure, government organized teams that fast reaction to evacuate floods victims and distribute early warning information to areas that prone of flood[150]. Not only governments, mitigation also prepared by public transportation company[151] and General election commission to anticipation of flood on the election day[152]. Many local government's websites[153,154,155,156,157,158,159,160,161,162] ministry's websites[163,164,165,166] and company's website[167,168] informed about their aids to the flood's victims. Aids supply to the flood victims in Indonesia worked well[169,170,171,172]. In some cases, victims complained for the uneven distribution of aids[173], lack of clean water and electricity[174], health problems[175,176], and some flooded areas has become worse than before[177].

5.4.4 Discussion of Flood in Indonesia

As the most frequent disasters in Indonesia, there are plenty of articles on online newspapers about flood in Indonesia. Not only the newspapers, but the official ministry's websites, locals' governments' websites also distributed the news about floods' victims. However, the websites publish different angels about disasters; most of the websites only publish articles about their institution distributing aids to the victims. It is more like the CSR reports. In case of flood, mass media also gives a good portion for the socialized flood mitigation, either the infrastructure that both locals and central government done or educating the public on how self-evacuated from flood. Online newspapers also suggest and inform the readers on how to do many mitigations, such as re-planting woods as water safer, checking the water tunnel regularly to make sure the water flows have no obstacles, etc. Regarding flood coming too often in Indonesia, we can get detailed information about the victims, number of people died, number of people evacuated, and even the numerous losses. In case of flood, only a few portions of drama are found at online newspapers. Aids distribution is still the highest issue of drama. Developing infrastructures and providing tools as flood mitigations by governments (both locals and central) can be seen as the realization of political commitment of the leaders to reduce the flood damages. So that, mass media has important role to report the leader's performance to the public. However, until now, there are none of the leaders that can make flood totally disappears either in the city or rural areas.

5.5 Discussion and Conclusion

Demands to meet the daily deadline force the disaster journalists to report un-double-checking data. As a result, the accuracy of the data is hard to get, for example: numbers of victim are different from one article to the other, so is the significant information about the magnitude power of the earthquake. However, information that connected to experts, informants, and related to time is mostly accurate. This relates to Sarma (2021) mentioned that sometimes mass media misunderstood role in disaster events. Moreover, mass media in Indonesia did their role in three dimensions that they were reported: impact, factual, and dramatization of the after disasters as Antunes, Pereira, Zêzere, and Oliveira (2022) mentioned.

(1) The impact of the mass media is mentioned in their articles: the losses suffered by the survivors include both losses their assets and even losses their members of families. Infrastructure damages as public losses are also reported by online newspapers. Online newspapers also reported the amount of money loss according to the government institution's reports. (2) For the factual: Online newspapers can be seen as the most update news as it can be read faster than conventional newspapers that update once a day. In case of disasters event, online newspapers update numbers of victims or found victims regularly in such short term of time. Moreover, factual of time of the hazard happened, the magnitude power, the process of evacuation, and damaged are reported by online newspaper. Somehow, (3) dramatization is important as commercial value for newspapers' industry. Therefore, stories of the survivors on the process of finding their members of families, corruptions, and unfair of aids distributions, have always a good portion in news.

On the role of disaster preparedness and management (mitigation), online newspapers did they role even though the portion needs to be enhanced. There are many ways online newspapers use to educate and inform the readers of mitigation. Interviews expert on how to handle and avoid the hazard is one of the tools online newspapers socialized of the mitigation. Highlighting the experiences and stories of the survivals of a disaster, which also contains information about what to do to survive a disaster. Online newspapers also report the infrastructure that government developed to reduce the risk of disasters. Repetitions of information to make sure that readers remember also mass media did such as information that high-power magnitude of earthquake brought the possibility of tsunami, or if the earthquake happens, they need to run into open spaces and avoid standing under the tree or pole. Mass media also suggests many tips to do small things to help mitigation: plan trees, keep water flow in the culverts smoothly, etc.

Online newspapers also play an important role of the social control in the disaster event. They report about aids distributions, and areas that lack aids, or uneven distribution of the aid. Moreover, they also confirmed about the aids' corruption or aids' looting. On the other hand, mass media reported the other impact of disaster to the social problem, such as published the peace MOU between GAM and Indonesia government. Role of mass media in terms of disaster also can be a reminder to the leaders about their political promises: What the leaders have done in terms of hazard and disaster mitigation.

Somehow, mass media also needs to fulfil their roles as companies. They need to bring income; therefore, the commercial values are needed. Eye-catching news such as articles about celebrities as survival of a disasters more interesting than local's survivals. Or facilitating big companies' CSR, or political party news about the aids to the victims in disasters. However, in high-tech eras, to promote and perform, many companies, parties, and stakeholders have their own websites. So that, even though the role of the online newspapers is needed by companies, and stakeholders to promote and show what they have done, it is not the only one tool for them.

To conclude, mass media in this term is the online newspapers on the role of disaster preparedness and disaster management, have done their functions: information, education, entertainment, and persuasive (Schramm 1964). Sure, at disaster issues no entertainment applied, so that in this case only information, education, and persuasive functions. Online newspapers can mention the most update news, since it is online based. However, it lacks of accuracy due to expectation of fast update. So, in many cases different article will have different numbers even though they discussed the same issue. Online newspapers have important role on the developing of disaster awareness as well. By giving space to the survivors to share their sad experiences, it aims not only generate readers' empathy but also develop the awareness of hazards and disasters by knowing the natural signs. Interviewed with expert to socializing reliable information to increase public's awareness. By increasing public awareness on hazards and disasters, it can be persuasive enough to the mitigation action much easier. Many online newspapers articles found suggested readers to do small things as mitigation.

Qualitative content analysis in this paper tries to compile the online newspaper articles by using keywords of name of disasters, places, and times. We found that information purposes are the most functions that newspapers did in terms of disasters: number of victims, times, and affected areas of disasters. Other than that, online newspapers are also frequently seen as the social control especially about international aids, local aids, aids distribution, and mitigation infrastructure built by governments as political promises. However, the expectation of "mass media can raise public awareness to the hazard and natural disaster (Khumairoh et al. 2021)" cannot be seen in this paper, because mass media is not the only tool to raise the public awareness even though they have important role of it.

Notes

1. https://news.detik.com/berita/d-5902060/proses-terjadinya-tsunami-dan-apa-itu-tsunami-simak-penjelasan-lengkapnya.
2. https://www.liputan6.com/global/read/3667115/menguak-asal-muasal-istilah-tsunami-dan-makna-di-baliknya?page=2.
3. https://sains.kompas.com/read/2014/12/24/18281591/.Perjalanan.Nama.Tsunami.

4. https://www.rctiplus.com/news/detail/travel/3462687/pernah-terjadi-di-indonesia--ini-8-tsunami-terbesar-dalam-sejarah.
5. https://nasional.sindonews.com/read/969311/15/tsunami-besar-yang-pernah-melanda-indonesia-nomor-4-menewaskan-lebih-dari-230-ribu-orang-1671023542/10.
6. https://www.cnbcindonesia.com/research/20230426074723-128-432271/10-tsunami-terbesar-di-ri-korban-jiwa-hingga-ratusan-ribu.
7. https://www.kompas.com/tren/read/2021/12/22/091202565/hari-ini-dalam-sejarah-tsunami-selat-sunda-menewaskan-437-orang?page=all.
8. https://www.detik.com/edu/detikpedia/d-6771006/10-tsunami-terbesar-di-dunia-dengan-korban-terbanyak-2-terjadi-indonesia.
9. https://nasional.sindonews.com/read/969311/15/tsunami-besar-yang-pernah-melanda-indonesia-nomor-4-menewaskan-lebih-dari-230-ribu-orang-1671023542/10.
10. https://www.bbc.com/indonesia/indonesia-45742383.
11. https://www.cnbcindonesia.com/research/20230426074723-128-432271/10-tsunami-terbesar-di-ri-korban-jiwa-hingga-ratusan-ribu.
12. https://www.kompas.com/tren/read/2021/12/22/091202565/hari-ini-dalam-sejarah-tsunami-selat-sunda-menewaskan-437-orang?page=all.
13. https://www.detik.com/edu/detikpedia/d-6771006/10-tsunami-terbesar-di-dunia-dengan-korban-terbanyak-2-terjadi-indonesia.
14. https://nasional.sindonews.com/read/969311/15/tsunami-besar-yang-pernah-melanda-indonesia-nomor-4-menewaskan-lebih-dari-230-ribu-orang-1671023542/10.
15. https://voi.id/memori/89320/guncangan-gempa-di-palu-dan-donggala-yang-memicu-tsunami-dan-likuifaksi-dalam-sejarah-hari-ini-28-september-2018.
16. https://nasional.tempo.co/read/1672427/hari-memilukan-bencana-tsunami-aceh-26-desember-2004-dalam-waktu-6-menit-habis-semua.
17. https://www.liputan6.com/news/read/93302/korban-tewas-tsunami-aceh-sumut-mencapai-101318-jiwa.
18. https://www.nu.or.id/warta/jumlah-korban-tewas-akibat-gempa-dan-tsunami-di-aceh-dan-sumut-mencapai-166080-orang-1trnp.
19. https://news.detik.com/berita/d-269632/korban-meninggal-akibat-tsunami-asia-mendekati-160-000-jiwa.
20. https://www.liputan6.com/hot/read/5042822/penyebab-tsunami-aceh-2004-bencana-alam-terparah-dalam-sejarah-modern.
21. https://aceh.tribunnews.com/2022/12/26/mengenang-18-tahun-tsunami-aceh-bencana-terbesar-dan-memori-kelam-26-desember-2004.
22. https://harianrakyataceh.com/2017/12/26/ustad-faizal-tsunami-bukan-hal-baru-di-aceh/.
23. https://www.liputan6.com/regional/read/5097293/mengenal-smong-mitigasi-bencana-ala-kearifan-lokal-simeulueaceh#:~:text=Dikutip%20dari%20dishub.acehprov.go,bencana%20gempa%20bumi%20dan%20tsunami.
24. https://nasional.kontan.co.id/news/tsunami-aceh-bencana-alam-terbesar-16-tahun-lalu.
25. https://www.detik.com/sumut/berita/d-6481655/7-fakta-dahsyatnya-tsunami-aceh-18-tahun-lalu.
26. https://nasional.tempo.co/read/1672427/hari-memilukan-bencana-tsunami-aceh-26-desember-2004-dalam-waktu-6-menit-habis-semua.
27. https://www.dw.com/id/kronologi-bencana-tsunami-2004-di-aceh/a-18146413.
28. https://tirto.id/sejarah-tsunami-aceh-2004-penyebab-dan-dampaknya-gBig.
29. https://www.detik.com/edu/detikpedia/d-6481762/detik-detik-tsunami-aceh-18-tahun-lalu-diawali-gempa-9-3-sr.
30. https://www.merdeka.com/trending/merinding-cerita-prajurit-tni-ini-selamat-dari-tsunami-aceh-nyangkut-di-tower.html.
31. https://timesindonesia.co.id/peristiwa/441310/kisahkisah-haru-tsunami-aceh-18-tahun-silam.

32. https://www.kominfo.go.id/content/detail/33511/misinformasi-tsunami-aceh-2004-disebabkan-oleh-ledakan-nuklir/0/laporan_isu_hoaks.
33. https://acehekspres.com/news/dampak-tsunami-aceh-terhadap-perdamaian-gam-dengan-ri/index.html.
34. https://www.liputan6.com/news/read/4442874/sederet-cerita-haru-dari-korban-tsunami-aceh-setelah-16-tahun-berlalu?page=3.
35. https://blogs.icrc.org/indonesia/bantuan-icrc-bagi-para-korban-gempa-dan-tsunami-aceh-10-tahun-lalu/.
36. https://news.detik.com/berita/d-336432/wfp-akan-salurkan-1-000-mt-bahan-pangan-bagi-korban-gempa.
37. https://www.indopos.co.id/nasional/2023/09/14/negara-pertama-bantu-tsunami-aceh-gfi-gempa-maroko-momentum-perkuat-solidaritas/.
38. https://nasional.tempo.co/read/54018/18-negara-berikan-bantuan-ke-aceh.
39. https://news.detik.com/berita/d-270876/17-negara-arab-beri-bantuan-jangka-panjang-korban-tsunami.
40. https://www.bbc.com/indonesia/indonesia-50682475.
41. https://www.ombudsman.go.id/perwakilan/news/r/pwk--16-tahun-tsunami-aceh-masih-ada-korban-menanti-bantuan-rumah.
42. https://www.bbc.com/indonesia/indonesia-50682475.
43. https://www.kompas.id/baca/nusantara/2022/12/26/18-tahun-tsunami-aceh-perkuat-kembali-edukasi-mitigasi.
44. https://www.voaindonesia.com/a/kisah-para-pejuang-mitigasi-bencana-dari-olokan-sampai-dianggap-sok-jadi-tuhan-/6976864.html.
45. https://www.borneonews.co.id/berita/286936-pakar-pengetahuan-mitigasi-bencana-tsunami-di-aceh-masih-minim.
46. https://www.antaranews.com/berita/3320607/pakar-pengetahuan-mitigasi-bencana-tsunami-di-aceh-masih-minim.
47. https://www.kompas.com/tren/read/2021/07/17/085500365/hari-ini-dalam-sejarah--gempa-dan-tsunami-pangandaran-668-tewas?page=all.
48. https://www.antaranews.com/berita/38162/proses-evakuasi-korban-tsunami-di-pangandaran-berlanjut.
49. https://www.sumenepkab.go.id/berita/baca/gempa-bumi-dan-tsunami-pangandaran-korban-tewas-menjadi-172-orang.
50. https://www.kompas.com/tren/read/2021/07/17/085500365/hari-ini-dalam-sejarah--gempa-dan-tsunami-pangandaran-668-tewas?page=all&lgn_method=google.
51. https://news.detik.com/berita-jawa-barat/d-5189821/cerita-mencekam-korban-selamat-tsunami-pangandaran-2006-lalu.
52. https://radartasik.id/mengintip-bulak-setra-pemukiman-yang-hilang-karena-tsunami-pangandaran-tahun-2006/.
53. https://www.harapanrakyat.com/2016/07/mengenang-tsunami-pangandaran-patimah-dapat-pertanda-sebelum-bencana/.
54. https://www.insiden24.com/ragam/pr-3965335820/tsunami-pangandaran-2006-kesaksian-abah-kentung-muncul-banyak-kejadian-ghaib-sebelum-dan-sesudah-tsunami.
55. https://news.detik.com/berita/d-638756/distribusi-bantuan-tsunami-pangandaran-melalui-kecamatan.
56. https://www.antaranews.com/berita/38258/adb-tawarkan-bantuan-pada-indonesia-untuk-atasi-tsunami-pangandaran.
57. https://www.antaranews.com/berita/38320/annan-pbb-siap-beri-bantuan-korban-tsunami-pangandaran.
58. https://regional.kompas.com/read/2008/11/19/12120210/~Regional~Jawa.
59. https://news.detik.com/berita/d-1488807/pemerintah-diminta-waspadai-korupsi-dana-bencana.
60. https://www.detik.com/jabar/berita/d-6298164/unesco-soroti-mitigasi-bencana-dan-jalur-evakuasi-tsunami-di-pangandaran.
61. https://rejabar.republika.co.id/berita/r0ahpv384/mitigasi-tsunami-pangandaran-butuh-breakwater.

62. https://www.bbc.com/indonesia/indonesia-45832237.
63. https://databoks.katadata.co.id/datapublish/2018/10/09/1948-jiwa-meninggal-akibat-gempa-dan-tsunami-palu-dan-donggala.
64. https://voi.id/memori/89320/guncangan-gempa-di-palu-dan-donggala-yang-memicu-tsunami-dan-likuifaksi-dalam-sejarah-hari-ini-28-september-2018.
65. https://news.detik.com/berita/d-4361370/sore-di-palu-dan-donggala-gempa-tsunami-dan-likuifaksi.
66. https://www.metrosulteng.com/sosial-budaya/pr-5194940624/tsunami-dan-likuifaksi-palu-bencana-langka-di-dunia-laut-dan-perut-bumi-mengamuk-alfatihah-untuk-korban.
67. https://news.republika.co.id/berita/pfy47b409/fenomena-likuifaksi-dan-tenggelamnya-rumahrumah-di-petobo.
68. https://www.bbc.com/indonesia/indonesia-45832237.
69. https://www.bbc.com/indonesia/indonesia-45795667.
70. https://www.bbc.com/indonesia/trensosial-45797943.
71. https://nasional.kompas.com/read/2018/10/01/19384081/ini-daftar-18-negara-yang-tawarkan-bantuan-untuk-palu-dan-donggala?page=all.
72. https://www.voaindonesia.com/a/negara-siap-bantu-korban-gempa-dan-tsunami-di-sulawesi-tengah/4599739.html.
https://www.cnbcindonesia.com/news/20181002163332-4-35740/14-negara-bantu-tangani-bencana-palu-donggala-ini-daftarnya.
73. https://nasional.tempo.co/read/1131926/18-negara-bantu-korban-gempa-donggala-dan-tsunami-palu.
74. https://www.voaindonesia.com/a/bantuan-asing-untuk-palu-telah-mencapai-220-miliar-rupiah-/4602407.html.
75. https://www.bbc.com/indonesia/indonesia-45824299.
76. https://www.bbc.com/indonesia/indonesia-45716700.
77. https://www.antaranews.com/berita/753065/bnpb-kemampuan-mitigasi-palu-donggala-masih-minim.
78. https://www.liputan6.com/showbiz/read/4937043/ifan-seventeen-sempat-ajak-bicara-jenazah-dylan-sahara-pasca-tsunami-disangka-gila-tapi-alasannya-bikin-pilu?page=4.
79. https://bangka.tribunnews.com/2018/12/25/inilah-daftar-nama-5-artis-yang-menjadi-korban-tsunami-di-banten?page=all.
80. https://www.bbc.com/indonesia/live/indonesia-46663949.
81. https://www.cnnindonesia.com/teknologi/20181231002758-199-357400/bmkg-paparkan-kronologi-tsunami-selat-sunda.
82. https://www.bbc.com/indonesia/indonesia-46674630.
83. https://elshinta.com/news/254309/2021/12/22/22-desember-2018-tsunami-terjang-pantai-banten-hingga-lampung.
84. https://www.liputan6.com/health/read/3857719/kemensos-kirim-bantuan-tanggap-darurat-tsunami-selat-sunda-ke-banten-dan-lampung.
85. https://poskota.co.id/2018/12/23/pemkot-tangerang-kirim-bantuan-ke-anyer-bantu-korban-tsunami-selat-sunda.
86. https://news.detik.com/berita/d-4360619/pemerintah-gelontorkan-rp-2-m-untuk-korban-tsunami-selat-sunda.
87. https://www.liputan6.com/regional/read/4438538/jejak-2-tahun-tsunami-selat-sunda-dan-upaya-penanggulangan-bencana-di-pesisir?page=2.
88. https://travel.detik.com/travel-news/d-4946616/pandemi-corona-buat-pariwisata-anyer-dan-carita-terpukul-dua-kali.
89. https://banten.suara.com/read/2020/05/26/143949/pusing-7-keliling-pebisnis-di-pantai-anyer-digempur-tsunami-sampai-corona.
90. https://www.merdeka.com/jabar/peristiwa-27-mei-2006-gempa-di-yogyakarta-yang-menewaskan-ribuan-orang-kln.html.
91. https://www.kompas.id/baca/foto/2023/05/25/mengenang-dahsyatnya-gempa-yogya-17-tahun-silam.

92. https://halosemarang.id/mengenang-gempa-bumi-yogyakarta-2006-ahli-tekankan-pentingnya-mitigasi-dan-mikrozonasi.
93. https://www.voaindonesia.com/a/lima-belas-tahun-gempa-yogya-57-detik-pelajaran-bagi-dunia/5906542.html.
94. https://www.krjogja.com/yogyakarta/1242473388/16-tahun-yang-lalu-gempa-bumi-besar-terjadi-di-yogyakarta.
95. https://www.liputan6.com/bisnis/read/5333595/fakta-fakta-gempa-bantul-yogyakarta-magnitudo-64-penyebab-dampak-terasa-hingga-jawa-timur?page=2.
96. https://www.kompas.com/sains/read/2021/05/27/120200623/mengenang-gempa-jogja-2006-ini-mekanisme-dan-potensi-bahayanya?page=all.
97. https://elshinta.com/news/303823/2023/05/27/27-mei-2006-gempa-m-59-guncang-yogyakarta.
98. https://jogja.tribunnews.com/2023/05/26/gempa-gempa-jogja-yang-tercatat-sejarah-paling-mematikan-sabtu-wage-27-mei-2006.
99. https://news.detik.com/berita/d-604871/34-negara-bantu-korban-gempa-yogya-jateng.
100. https://www.voaindonesia.com/a/a-32-2006-05-31-voa2-85385782/60743.html.
101. https://news.detik.com/berita/d-687406/dana-rekonstruksi-korban-gempa-yogya-cair-5-oktober.
102. https://news.detik.com/berita/d-741603/turki-bangun-100-rumah-untuk-korban-gempa-di-bantul.
103. https://www.voaindonesia.com/a/jrf-124718879/94957.html.
104. https://news.republika.co.id/berita/qhh45t396/mengenang-gempa-padang-11-tahun-lalu.
105. https://voi.id/memori/90150/gempa-padang-meluluhlantakkan-pemukiman-dalam-sejarah-hari-ini-30-september-2009.
106. https://langgam.id/gempa-sumbar-2009-pengalaman-tanpa-pembelajaran/.
107. https://voi.id/memori/90150/gempa-padang-meluluhlantakkan-pemukiman-dalam-sejarah-hari-ini-30-september-2009.
108. https://www.liputan6.com/global/read/3112264/30-9-2009-gempa-76-sr-mengguncang-padang-sumatera-barat.
109. https://langgam.id/gempa-sumbar-2009-pengalaman-tanpa-pembelajaran/.
110. https://voi.id/memori/90150/gempa-padang-meluluhlantakkan-pemukiman-dalam-sejarah-hari-ini-30-september-2009.
111. https://news.detik.com/berita/d-1213784/negara-asing-berbondong-bondong-bantu-indonesia.
112. https://www.liputan6.com/news/read/246478/gempa-dan-bantuan-yang-tersendat.
113. https://nasional.tempo.co/read/278510/sumatera-barat-mendapat-rp-2-triliun-untuk-rumah-korban-gempa.
114. https://www.antaranews.com/berita/156561/bantuan-gempa-sumbar-dijarah.
115. https://tekno.tempo.co/read/1790836/cianjur-masuk-daerah-rawan-gempa-ini-penjelasan-badan-geologi.
116. https://www.bbc.com/indonesia/articles/c84p7znl7plo.
117. https://www.cnbcindonesia.com/news/20221122155219-4-390239/bukan-sekali-cianjur-langganan-gempa-sejak-zaman-belanda.
118. https://www.bbc.com/indonesia/articles/crgz23p0v60o.
119. https://news.republika.co.id/berita/rmyrgb368/korban-meninggal-dunia-akibat-gempa-cianjur-602-jiwa#:~:text=BNPB%20telah%20memverifikasi%20jumlah%20korban%20meninggal%20dunia%20sebanyak%20602%20jiwa.&text=EPA%2DEFE%2FMAST%20IRHAM%20Tim,dipastikan%20bertambah%20menjadi%20602%20jiwa.
120. https://news.detik.com/berita/d-6418520/tentang-sesar-cimandiri-diduga-terkait-gempa-cianjur-21-november.
121. https://news.detik.com/berita/d-6418520/tentang-sesar-cimandiri-diduga-terkait-gempa-cianjur-21-november.
122. https://jabarprov.go.id/berita/penyaluran-dana-stimulan-korban-gempa-cianjur-target-selesai-akhir-2023-8946.

123. https://www.pasuruankab.go.id/isiberita/peduli-korban-gempa-cianjur-pemkab-pasuruan-salurkan-bantuan-rp-70-juta.
124. https://dprd.bandungbaratkab.go.id/asdeksi-jabar-salurkan-bantuan-untuk-korban-gempa-bumi-di-cianjur/.
125. https://cilacapkab.go.id/v3/bpbd-cilacap-salurkan-bantuan-donasi-untuk-korban-gempa-cianjur/.
126. https://csr.jabarprov.go.id/post/penerimaan-bantuan-untuk-gempa-cianjur-oleh-gubernur-jawa-barat-dari-hyundai-group.
127. https://www.bekasikab.go.id/camat-se-kabupaten-bekasi-berikan-bantuan-korban-gempa-cianjur.
128. https://www.daya-group.co.id/article/daya-group-salurkan-bantuan-kepada-korban-gempa-bumi-di-cianjur.
129. https://www.pertamina.com/id/news-room/news-release/pertamina-gerak-cepat-salurkan-bantuan-sembako-hingga-kirim-tim-medis-untuk-korban-gempa-cianjur.
130. https://nasional.tempo.co/read/1662819/bupati-sayangkan-bantuan-korban-gempa-cianjur-tak-terkoordinasi-posko-utama.
131. https://www.merdeka.com/peristiwa/banjir-jadi-bencana-paling-sering-terjadi-di-indonesia-ini-cara-penanggulangannya.html.
132. https://sultimnews.info/2022/09/05/terdampak-banjir-kades-indang-sari-sebut-bukan-murni-bencana-alam-box-kulvert-penyebab-sebenarnya-air-meluap/.
133. https://kalbar.prokal.co/read/news/5768-banjir-di-kalbar-bukan-murni-bencana-alam.
134. https://www.bbc.com/indonesia/articles/cgrymv8lj8lo.
135. https://www.bbc.com/indonesia/indonesia-50982184.
136. https://www.detik.com/jateng/berita/d-6510107/bpbd-jateng-kerugian-akibat-banjir-awal-2023-capai-rp-16-m.
137. https://www.kompas.id/baca/nusantara/2022/10/11/kerugian-akibat-banjir-di-aceh-ditaksir-hingga-puluhan-miliaran-rupiah.
138. https://www.viva.co.id/berita/nasional/1090275-daftar-20-kota-rawan-banjir-di-indonesia?page=1.
139. https://metro.sindonews.com/read/1035191/170/banjir-jakarta-terparah-sejak-1996-2020-nomor-4-tewaskan-80-orang-1677647015.
140. https://www.suara.com/news/2022/07/17/053555/6-banjir-terbesar-sepanjang-sejarah-di-indonesia-ratusan-nyawa-melayang.
141. https://www.beritasatu.com/news/866761/6-banjir-terbesar-sepanjang-sejarah-di-indonesia.
142. https://sains.sindonews.com/read/1205859/766/7-banjir-bandang-terbesar-yang-pernah-melanda-indonesia-1695186558.
143. https://www.detik.com/edu/detikpedia/d-6182544/5-dampak-banjir-bagi-masyarakat-dan-lingkungan.
144. https://www.prudential.co.id/id/pulse/article/5-dampak-banjir-bagi-masyarakat-yang-perlu-diketahui/.
145. https://news.detik.com/berita/d-6574137/mitigasi-bencana-banjir-sebelum-saat-dan-sesudah-banjir-simak-di-sini.
146. https://www.antaranews.com/berita/3187317/pakar-mitigasi-banjir-perlu-didukung-perluasan-penyebaran-informasi.
147. https://banten.antaranews.com/berita/269838/pemkot-tangerang-jakbar-perkuat-koordinasi-soal-penanganan-banjir.
148. https://www.cnnindonesia.com/nasional/20231025180231-25-1015926/pemkot-semarang-perkuat-mitigasi-banjir-jelang-musim-hujan.
149. https://www.kompas.id/baca/metro/2023/11/26/kesiagapan-pompa-air-di-titik-rawan-banjir-jakarta-diperlukan.
150. https://www.jawapos.com/nasional/013233599/mitigasi-banjir-bpdb-dki-jakarta-bentuk-267-personil-tim-reaksi-cepat.
151. https://jakartamrt.co.id/id/info-terkini/mitigasi-ancaman-banjir-di-area-stasiun-mrt-jakarta.
152. https://www.kompas.id/baca/metro/2023/12/07/kpu-dki-jakarta-mitigasi-2841-tps-rawan-banjir.

153. https://dinsos.bulelengkab.go.id/informasi/detail/artikel/bantuan-pertolongan-yang-dibutuhkan-korban-bencana-alam-tragedi-kemanusiaan-61.
154. https://bpbd.bojonegorokab.go.id/berita/baca/117.
155. https://baitulmal.acehprov.go.id/bantuan-untuk-korban-bencana-alam-bencana-kemanusiaan-dan-musibah-bencana-lainnya.
156. https://dinsos.kaltimprov.go.id/pemprov-kaltim-beri-bantuan-kepada-korban-banjir-di-long-kali/.
157. https://pariamankota.go.id/berita/wako-genius-umar-serahkan-bantuan-untuk-korban-banjir.
158. https://acehbaratkab.go.id/berita/kategori/berita/dinsos-dan-bpbd-salurkan-bantuan-masa-panik-korban-banjir-banjir.
159. https://karawangkab.go.id/berita/bjb-karawang-berikan-bantuan-untuk-korban-banjir-1.
160. https://dinsos.labura.go.id/post/lihat/PEMBERIAN-BANTUAN-SEMBAKO-KEPADA-KORBAN-BENCANA-BANJIR-DI-DESA-LOBU-RAMPAH-KECAMATAN-MARBAU-KABUPATEN-LABUHANBATU-UTARA.
161. https://diskominfo.kaltaraprov.go.id/salurkan-bantuan-ke-korban-banjir/.
162. https://jatengprov.go.id/publik/korban-banjir-dapat-bantuan-logistik-dari-pemprov-jateng/.
163. https://pu.go.id/berita/penyaluran-bantuan-bagi-korban-bencana-banjir-di-nad.
164. https://www.kominfo.go.id/content/detail/23704/kominfo-salurkan-bantuan-untuk-korban-banjir-jabodetabek/0/sorotan_media.
165. https://ksdae.menlhk.go.id/artikel/11466/solidaritas-meringankan-beban-korban-banjir.html.
166. https://www.djkn.kemenkeu.go.id/kpknl-palu/baca-berita/29589/KPKNL-Palu-Salurkan-Bantuan-kepada-Korban-Banjir-Bandang-Sigi.html.
167. https://alfamart.co.id/news-detail/alfamart-alirkan-bantuan-untuk-korban-banjir.
168. https://www.temposcangroup.com/id/read/Dukung-Pemulihan-Korban-Banjir-Tempo-Scan-Group-Gelar-Program-Bakti-Sosial-Bantuan-Bencana-Banjir-Jakarta-2020.
169. https://www.antaranews.com/berita/3372837/pemkab-kupang-distribusi-bantuan-bagi-korban-bencana-banjir-niunbaun.
170. https://www.antarafoto.com/view/1840759/distribusi-bantuan-sosial-korban-banjir-di-kalteng.
171. https://www.antaranews.com/berita/3891354/riau-salurkan-bantuan-logistik-tahap-dua-untuk-korban-banjir-rohul.
172. https://megapolitan.kompas.com/read/2014/02/10/0720091/Bagaimana.Mekanisme.Penyaluran.Bantuan.Korban.Banjir.di.Jakarta.?page=all.
173. https://www.kompas.id/baca/nusantara/2021/01/20/distribusi-bantuan-logistik-korban-banjir-kalsel-belum-merata.
174. https://sumsel.inews.id/berita/keluhan-korban-banjir-di-lahat-kesulitan-air-bersih-dan-listrik-masih-padam.
175. https://www.radarbogor.id/2021/01/24/ratusan-warga-korban-banjir-bandang-puncak-alami-keluhan-kesehatan/.
176. https://www.antarafoto.com/id/view/1560925/keluhan-kesehatan-korban-banjir-lombok.
177. https://www.cnnindonesia.com/nasional/20200225125930-20-477867/keluhan-warga-ketika-banjir-melanda.

References

Antunes MN, Pereira SDS, Zêzere JL, Oliveira AE (2022) Disaster journalism in print media: analysis of the top 10 hydrogeomorphological disaster events in Portugal, 1865–2015. Int J Disaster Risk Sci 13:521–535. https://doi.org/10.1007/s13753-022-00425-2

Bansal S (2018) Role of media in disaster management. Int J Creat Res Thoughts (IJCRT) 6(2) | ISSN: 2320-2882

Bengtsson M (2016) How to plan and perform a qualitative study using content analysis. NursingPlus Open 2:8–16. https://doi.org/10.1016/j.npls.2016.01.001

BMKG (2023) Latest Earthquakes (< 5). https://www.bmkg.go.id/gempabumi-terkini.html

Damayanti QV, Suni AMF, Rabbani H, Hayuningfitriaya NG, Rahayu SA, Asmara G (2021) Mitigasi Bencana dan Peran Kearifan Lokal Melalui Sesar Palukoro dan Sesar Lembang. Cinematol J Anthol Film Televis Stud 1(3):1

Donohue GA, Tichenor PJ, Olien CN (1973) Mass media functions, knowledge and social control. J Quart. https://doi.org/10.1177/107769907305000404

Ermanto (2005) Menjadi WARTAWAN Handal & Profesional (Becoming a Reliable and Professional Journalist). Yogyakarta: Cinta Pena

Fauzi D, Marhamah D (2020) The pattern of disaster communication and media to improve community alertness in North Aceh regency. J Pekommas 5(2):133–142. https://doi.org/10.30818/jpkm.2020.2050203

Khumairoh Z, Widana IDKK, Sumantri SH (2021) The role of communication as the disaster risk reduction in Indonesia capital city transference policy. In: 2nd international conference on disaster and management IOP conf. ser.: earth environ. sci., vol 708, p 012101

Krippendor K (2004) Content analysis an introduction to its methodology, 2nd edn. University of Pennsylvania Sage Publications Inc., Thousand Oaks

Matthews J (2017) The role of a local newspaper after disaster: an intrinsic case study of Ishinomaki, Japan. Asian J Commun 27(5):464–479. https://doi.org/10.1080/01292986.2017.1280065

McCombs M, Valenzuela S (2007) The Agenda-Setting Theory" Cuadernos de Información, núm. 20, julio, 2007, pp 44–50 Pontificia Universidad Católica de Chile Santiago, Chile

Nahak FN (2021) Role of media in disaster preparedness: some case studies from calamity prone Odisha. E3S Web Conf 331:04010 (2021) ICDMM 2021. https://doi.org/10.1051/e3sconf/202133104010

Nasution SH (2018) Information Crowdsourcing: Utilizing ICT For Indonesia's Disaster Management System. Case study series #42 December 2018 Center for Digital Society, Faculty of Social and Political Sciences, Gadjah Mada University, Yogyakarta, Indonesia

Pratomo RA, Rudiarto I (2013) Permodelan Tsunami dan Implikasinya Terhadap Mitigasi Bencana di Kota Palu. Jurnal Pembangunan Wilayah dan Kota 9(2):174–182 Juni 2013 Biro Penerbit Planologi Undip

Sarma RJ (2021) Media and disaster management: the role of media. Int J Adv Eng Manag (IJAEM) 3(10):827–833. www.ijaem.net. ISSN: 2395-5252

Scanlon J (2011) Research about the mass media and disaster: never (well hardly ever) the twain shall meet" chapter at "Journalism: Theory and Practice" Jason R Detrani (Editor)

Schramm W (1964) Mass media and national development—the role of information in the developing countries. Stanford University Press, Stanford

Sukino WG, Samad MA, Mangngasing N, Rivai A (2019) Manajemen Mitigasi Bencana Kota Palu (Palu City Disaster Mitigation Management). J Public Admin Gov 1:1

Tekeli-Yesil S, Kaya M, Tanner M (2018) The role of the print media in earthquake risk communication: information available between 1996 and 2014 in Turkish newspapers. Int J Dis Risk Reduct 33:284–289

White MD, Marsh EE (2006) Content analysis: a flexible methodology. Lib Trends 55(1): Summer 2006 ("Research Methods," edited by Lynda M. Baker), pp. 22–45 © 2006 The Board of Trustees, University of Illinois

Is Flood Insurance Scheme in Regional Areas Feasible? An Explorative Study in Riau, Indonesia

6

Ikval Suardi, Felix Wisnu Handoyo, Chitra Indah Yuliana, Syarifah Aini Dalimunthe, Erla Mychelisda, Purwanto Purwanto, Irwanda Wisnu Wardhana, and Agus Eko Nugroho

Abstract

Floods are among the most common and devastating natural hazard-induced disasters, particularly in vulnerable regional areas. This exploratory study examines the introduction of a flood insurance scheme in regional contexts, focusing on the case of Kampar District, Riau, Indonesia. It will explore the socioeconomic and geographical factors influencing the potential success of an insurance scheme, taking into account the characteristics of the community. Utilizing a mixed-methods approach, including quantitative data analysis from questionnaire and qualitative interviews with stakeholders, the research investigates the current state of local government's flood risk management, the community awareness and preparedness, and the business related to risk disaster's insurance. The study reveals the challenges and opportunities associated with the feasibility in implementing a flood insurance scheme, considering factors such as disaster risk perception, institutional capacity, community engagement, and willingness to pay in insurance scheme. Findings suggest that while certain barriers exist, such as limited financial resources and lack of awareness, there is potential for improving the role of local stakeholders for a well-designed flood insurance scheme to enhance regional resilience and recovery. The study contributes to the broader discourse on disaster risk reduction and mitigating the potential impact of floods in developing regions. Further research and collaboration are recommended to refine the proposed flood insurance framework and address the identified obstacles for effective implementation in regional settings.

Keywords

Flood insurance scheme · Regional resilience · Disaster risk management · Community preparedness · Socioeconomic factors

I. Suardi · F. W. Handoyo (✉) · C. I. Yuliana · P. Purwanto
Research Center for Behavioral and Circular Economics, National Research and Innovation Agency, Jakarta, Indonesia
e-mail: felixwisnuhandoyo@gmail.com

S. A. Dalimunthe
Research Center for Population, National Research and Innovation Agency, Jakarta, Indonesia

E. Mychelisda · A. E. Nugroho
Research Center for Macroeconomics and Finance, National Research and Innovation Agency, Jakarta, Indonesia

I. W. Wardhana
Research Center for Cooperative, Corporation, and People's Economy, National Research and Innovation Agency, Jakarta, Indonesia

D. Hidayati et al. (eds.), *Examining Disaster Risk Reduction in Indonesia*, Sustainable Development Goals Series, https://doi.org/10.1007/978-981-96-4182-6_6

6.1 Introduction

Climate change is poised to exert a significant impact on water-related disasters and environmental challenges in Indonesia, particularly in regions experiencing an anticipated rise in extreme monsoon-related rainfall. This increase in rainfall is expected to elevate the vulnerability of a larger segment of the population to floods (Marfai et al. 2015; Yamamoto et al. 2021). The focus of our study is on the Kampar Watershed, situated in Sumatra Island, where coastal lowland areas primarily consist of peatland.

Peatland, a wetland environment vital for its ecological balance, has historically relied on natural flooding from rivers. However, recent years have witnessed extensive development in these areas, primarily for oil palm and other plantations. A previous study shows the alterations in the frequency and magnitude of flood inundation, attributed to climate change, pose a threat to the delicate peatland environment and its carbon stores. Moreover, these changes may damage the agricultural plantations in the Sumatra and Kalimantan Islands of Indonesia, particularly in the Kampar Watershed region.

Current economic losses from weather-related hazards are substantial and anticipated to rise further. This increase is primarily attributed to socioeconomic developments and climate change, exposing valuable assets to heightened flood risks. The escalating negative impacts of flood losses present a challenge to governments, the public sector, and the private sector to formulate sustainable flood risk management mechanisms to mitigate these losses (Gomez-Cunya et al. 2022; Slovic and Peters 2006).

Recognizing the slow progress and limited success of climate change mitigation in reducing greenhouse gases, a shift toward sustainable adaptation is a crucial strategy for managing climate change risks. Adaptation involves adjusting to current or future implications of climate change to avoid harm or exploit benefits. Flood risk is intricately linked to the hazard itself, population, assets, and values at risk (exposure), as well as the vulnerability of society—its capacity to cope with flood events (Kron 2005). Flood risk reduction measures can be categorized based on their influence on these flood risk determinants. Additionally, these measures address floods of varying strength and return periods, emphasizing the need for tailored emergency management and adaptation strategies.

Uncontrolled land use changes and inadequate coordinated management upstream and downstream of river basin areas pose significant threats, leading to devastating floods. This chapter delves into the intricate interplay between flood dynamics and non-infrastructure mitigation against disaster, focusing on the Kampar Watershed in the Riau Province, Kampar District. The Kampar River, a vital watercourse in the region, spans approximately 413.5 km and serves as a lifeline for the communities in its vicinity. However, rampant land use changes, transforming catchment areas into residential, industrial, and commercial zones, as well as agricultural land, have significantly altered the hydrological dynamics of the Kampar Watershed. The convergence of two major rivers, Kampar Kanan and Kampar Kiri, forms the Kampar River, creating a complex network of tributaries, such as Sibayang and Singingi Rivers, which are extensively utilized by the local population for various purposes. The Kampar Watershed, a critical component of the Kampar district, is endowed with a vast forest resource covering 36,577.5 hectares, constituting 40% of the district's total area (Kausarian et al. 2023). The escalating land use changes in the Kampar Watershed have led to alterations in the runoff coefficient, resulting in increased surface runoff and influencing river discharge. Understanding the intricate relationship between land use changes and flood dynamics in this region necessitates a comprehensive study that considers both advanced mitigation from socioeconomic point of view.

As rivers play a pivotal role in sustaining life and livelihoods, the consequences of floods are far-reaching. While floods bring certain advantages, their adverse impacts, especially in densely populated floodplain areas, are dominant and can be catastrophic. The continuous growth of population, elevated living standards, urbanization, and industrialization in flood-prone areas contribute to societal vulnerability (Bubeck et al. 2012; Champonnois and Erdlenbruch 2021).

This chapter explores the role of flood insurance as a crucial tool in mitigating the impact of floods on communities. Unlike traditional flood measures that aim to reduce damages, flood insurance diffuses losses over time, providing financial support to affected individuals and communities (Surminski and Eldridge 2017; Tariq et al. 2014). The implementation of flood insurance is often supported by legal provisions, public incentives, and enforcement practices, emphasizing the importance of a multi-faceted approach to flood mitigation in the Kampar Watershed and similar regions globally.

This research uses a mixed-methods approach to comprehensively investigate the state of flood risk management within local government (Mendes-Da-Silva et al. 2021; Roder et al. 2019). Through quantitative analysis of community questionnaires and qualitative interviews with key stakeholders, the study delves into the effectiveness of current strategies, community awareness, preparedness, and insurance integration as a mitigation tool. By synthesizing data from both sources, the research endeavors to provide a nuanced understanding of the dynamics at play. Furthermore, it seeks to generate evidence-based policy recommendations to enhance overall resilience by incorporating flood insurance schemes.

The study focuses on Riau as a case study representing the developing country and on household insurance. The remainder of this chapter is structured as follows: The second section discusses the literature review on mitigating flood risk through flood insurance. The third section explains the research methodology, including the description of data collection and the analytical approach used. Section four presents the field survey, particularly the discussion on the Feasibility Assessment of willingness to pay for the flood insurance scheme. Section five focuses on conclusions and policy recommendations drawn from the study.

6.2 Literature Review on the Complexity of Building Resiliency on Flood Hazards

6.2.1 The Historical Specificity of Flood Hazards Linked to Climate Change, Deforestation, and Urbanization

Flooding naturally refers to a sudden increase in overflowing water beyond its regular conditions. It varies from relatively small and manageable to rapid and disastrous events impacting the livelihoods of affected regions. In terms of location, floods are often distinguished into river, coastal, urban, and flash floods (Glago 2021). River floods occur when natural events, particularly a long period of heavy rains, lead to a large volume of water influx exceeding the carrying capacity of rivers, hence inundates wide areas of dry lands, particularly along the lower basins of the rivers. Similarly, coastal floods occur when the substantial rising of sea level swamps wide areas beyond the coastline, while flash floods refer to quick-rising floods that often severely impact the affected regions (Glago 2021). Moreover, when the incidence of these floods inundates broad areas of cities, this refers to urban floods. Numerous preceding events, including long-heavy rains, dam failures, earthquake-induced tsunami, coastal storms, strong ocean waves, and cyclones can thus trigger the occurrence of such floods. In this regard, flood hazards are deemed to be the most prevalent catastrophe to human lives, causing widespread destruction of housing property, buildings, infrastructure, and farmlands and even fatalities throughout history and places (Angelakis et al. 2023; Ghapar et al. 2018; Glago 2019, 2021; Jevrejeva et al. 2018; Rentschler et al. 2022).

According to Rentschler et al (2022), approximately about 1.81 billion people (23% of the world population) have been directly prone to flood hazards. The global number of flood-exposed people has stood roughly at 39% above the threshold predicted by Jongman et al. (2012), amounting to about 1.3 billion people by 2050. However, the number of people exposed to flood hazards varies through time and regions as presented in Fig. 6.1. Here, two district features are worthwhile noticed. Firstly, the average number of flood-affected people worldwide remained relatively low during the period of 1900s–1950s which then remarkably increased during the period of 1960s–1990s. The number of flood-affected people, however, decreased accordingly since the 2000s–2020s. Secondly, in comparison with developed countries, less and developing nations have been highly vulnerable to flood hazards, particularly those in South and East Asia and Sub-Sahara Africa (Fig. 6.1).

The increasing risks of flood hazards in the regions are particularly associated with higher rates of poverty and rapid urbanization combined with insufficient infrastructure and strategy to control and mitigate the impacts of flood hazards (Rentschler et al. 2022). Furthermore, it is also increasingly recognized that the worldwide risks of flood hazards have been largely associated with current problems of climate change. As Angelakis et al. (2023) points out, extreme weathers linked to climate change currently have been seen to be the main cause of increasing global catastrophes, especially floods, coastal storms, and cyclones. This is particularly in parallel with increasing global precipitation anomalous along with the rising of climatic problems.

With regard to Fig. 6.2, in the period of 1900s–1960s, the impacts of flood to the world society remained significantly low in terms of affected people, individual displacement, flood victims with immediate support, and economic damages as percentage of the world GDP. The flood impacts to the society increased remarkably from the period of 1960s/70s–1990s. The reason for the rising number of flood-affected people was associated with intensifying occurrence of flood hazards, combined with weakness in flooding mitigation strategies and early warning systems, especially in response to rapid growing population and urbanization during the period. In the period of 2010s–2020s, the impacts of flood disaster to the world society then decreased considerably along with global improvement strategies in flood control and management, particularly through enhancing flood disaster risk awareness, preparedness, and mitigation in vulnerable areas (Glago 2019; Aerts et al. 2018).

However, Fig. 6.2c indicates that as percentage of the world GDP, the damaging impacts of flood disaster to the global economy have a tendency to increase from the period of 1960s–2020s. The industrialization process linked to the upscaling value of global economic activities and upsetting of infrastructure development, especially in urban areas have led the world economy to become sensitive toward the occurrence of floods especially in urban areas. Large-scale productions of agricultural commodities, manufacturing products, and mining sector have generated over exploitation of natural resources, leading to the continuing degradation of the environment. Consequently, not only is the urban environment, but also the human biosphere becomes increasingly prone to natural as well as man-made disasters such as floods, massive landslides, famine, pandemic-induced economic crises, and various climatic disasters, such as long drought, cyclones, and heavy storms.

In this regard, historical views of scrutinizing the diverging modes of global productions are critical because the long-wave patterns of climatic variation interplay with the process of ecological degradations which are greatly affected by the change and metamorphosis of the worldwide production styles (O'Hara 2023). Figure 6.3 shows three important features causing the rise of climatic variability through time. Firstly, the global temperature-induced climatic change begun to manifest along with the emerging modes of greater production scales since the industrialization era in the late-nineteenth century. The upward trend of global temperature became critically apparent during the period of 1970s–2010s onward in parallel with the working of agricultural and industrial productions at greater scales (IPCC 2021).

The global temperature stood approximately at about 1.07–1.2 °C above mean over the period

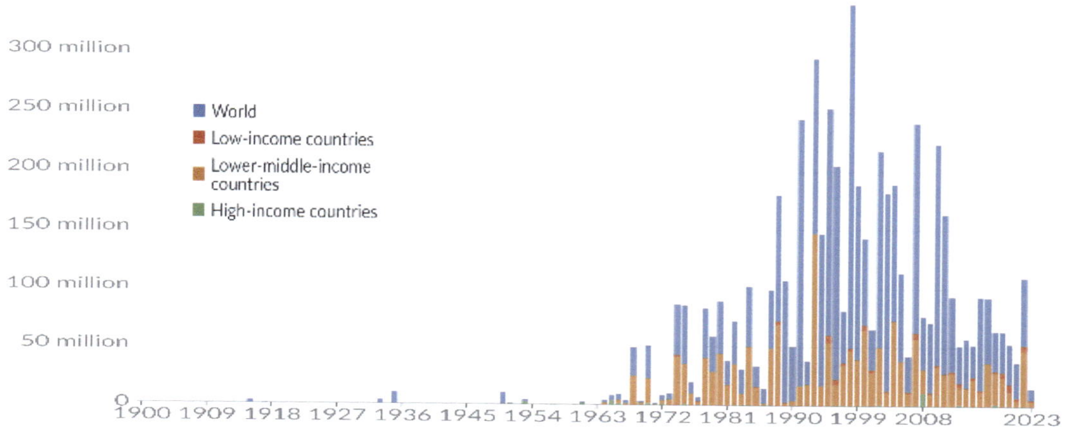

Fig. 6.1 Number of flood-affected people in the period of 1900–2023 [*Source* Our World in Data (n.d.) http: www.ourworldindata.org]

of 1900s–2020s which was significantly above the period before the 1900s, accounting for 0.4–0.9 °C below mean. Here, the rising of global temperature goes along with accelerating growth of energy production and consumption, especially coal and oil–gas fuel combustion, combined with massive exploitation of natural resources, land-use clearing, and deforestation. The ecological destructive forces lead to the abundant accumulation of CO_2 equivalent emissions[1] which then adversely impact the global warming. The global productivity increased significantly at the cost of biospheric destruction and the abundant accumulation of GHG (CO_2 equivalent) emission over time due to massive uses of fossil-fuel energy to support the working of global transportation systems, logistic distributions, building, and manufacturing, and farming productions (O'Hara 2023).

Secondly, the large scales of agricultural production and industrialization have been the major driving forces of flood hazards, especially through deforestation and urbanization, linked to abundant soil erosion and sedimentation of rivers. The global deforestation process has been associated with the massive production of timber and plywood for worldwide urban housing, infrastructure, and industrial constructions. In response to population growth, the increasing agricultural production has also led to farmland expansions at greater scales, such as through corporate plantation and pesticide and chemical fertilizer-based farming production. Such farming production styles further accelerate the process of deforestation, and GHC emissions, impacting global warming and, thus, climate change. This process has also been propelled by global mining activities often requiring massive land-use clearing.

In the context of flood-induced disasters, the uninterrupted process of deforestation significantly reduces the water absorption capability of soil and vegetation, leading to substantial runoff water influx directly to the rivers. As deforestation creates substantial sediment accumulations in rivers and floodplains, the problems of riverine sedimentation then cause a large volume of overflowing water to the dry lands. As a result, the rural inhabitants and farmland of the world, especially in developing countries across the downstream areas of river basins, become increasingly prone to flood hazards, particularly during heavy rains, frequent storms, and cyclones.

Thirdly, apart from facing many fatalities, housing displacement, and public infrastructure damage, the aftermath of flood hazards

[1] Agricultural activities, including livestock, land-use farming, chemical fertilizer and insecticide, agricultural and forestry pollution contributed about 20.1% of GHG emissions, energy use for industrial productions, pollution and waste and land-use clearing is about 29.4%, and energy use for transportation and supply chain contributing another 33.7% of GHG emissions (O'Hara 2023).

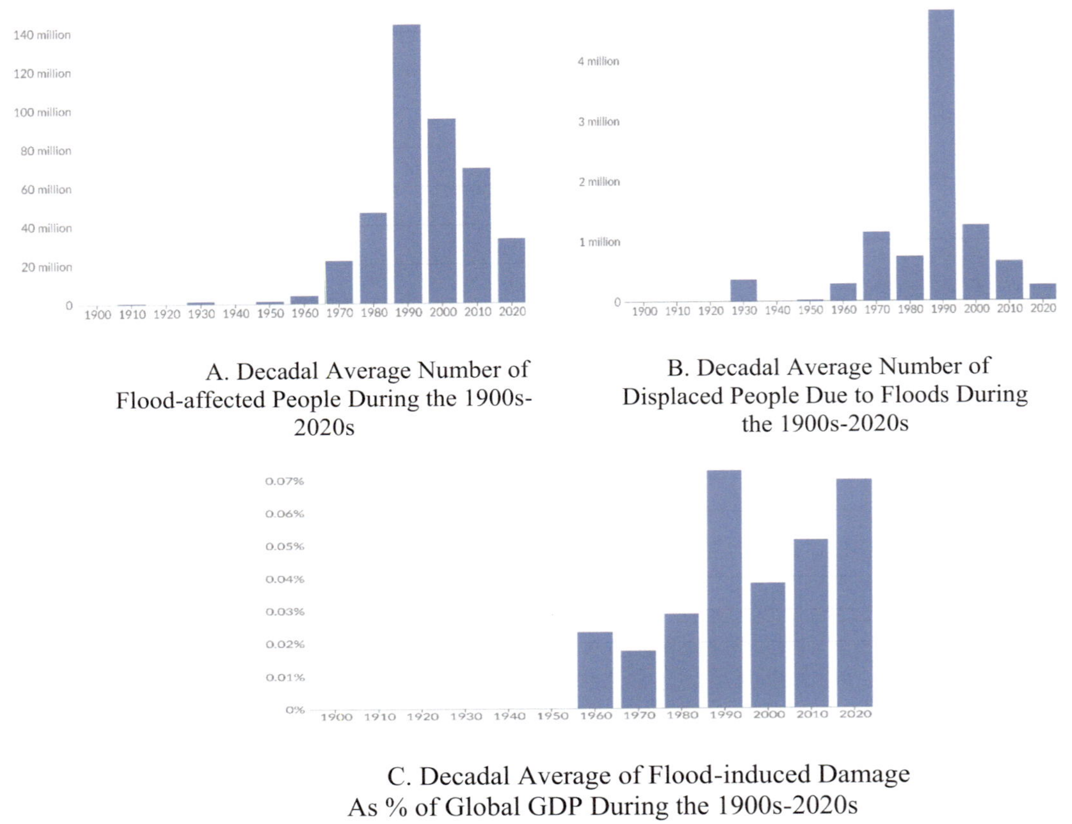

Fig. 6.2 **A** Decadal average number of flood-affected people during the 1900s–2020s. **B** Decadal average number of displaced people due to floods during the 1900s–2020s. **C** Decadal average of flood-induced damage as % of global GDP during the 1900s–2020s [*Source*: Our World in Data (n.d.) http: www.ourworldindata.org]

often results in serious public health problems, impacting the affected regions. Flood-induced disease epidemics, such as malaria, typhoid, and cholera, are typically common for less developed and developing countries in the tropical region following the occurrence of flood hazards. The impacts of devastating flood hazards also often create serious consequences to the affected individuals, such as increasing levels of emotional stress, extreme tiredness, and many other physiological health problems. Such intangible problems can be associated with numerous factors, including the death or serious injuries, displacements, living in overcrowded temporary accommodations, severe damage to homes and assets, loss of income sources, and the like. This is particularly the case for the poor and low-income families with insufficient savings, lack of disaster preparedness, and no insurance coverage (Glago 2021; Mwape 2009).

In the context of Indonesia, Table 6.1 presents an interconnected motion between the long-term dynamic of the economy and urbanization, linked to the increasing occurrence of flood hazards throughout 1900–2023. During the period of Dutch colonialization in 1900–1940, the major production styles of the Indonesian economy were characterized by the introduction of the Cultivation system that forced local farmers to grow world commodities, incorporated with large-scale productions of colonial plantations, especially spice products (e.g., nutmeg, pepper,

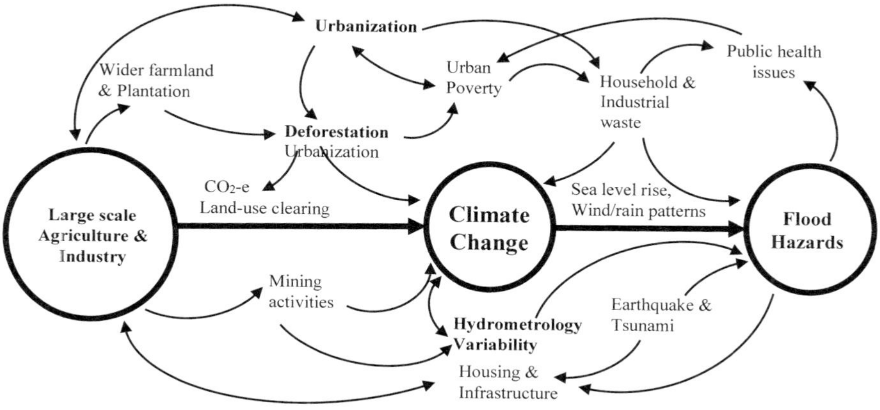

Fig. 6.3 The circular and cumulative causation of climate change affecting flood hazards [*Source* Adapted from O'Hara (2023)]

cloves), coffee, rubber, tea, sugar, and timber commodities (Booth 1998). The value of exports increased substantially during this period, yet the annual growth per capita remained relatively low, accounting for about 1.7%, particularly due to excessive labor exploitation and slavery. The growth of per capita income turned to − 3.4%, as the significant depression era of 1929–1940 dampened the world demand for major commodities of the colonial plantations.

Significant deforestation occurred mainly in Java as the centers of colonial administration, plantation and population. Approximately in the year 1900 about, 99.2% of the total land area of Indonesia was covered by forest, especially in Kalimantan, Sumatera and Sulawesi Mackinnon (1997), while the forest cover reduced to only about 1.3% by the late 1900s (GFW/FWI 2002). Having relatively good forest cover and low urbanization, the country had faced a low incidence of flood disasters. In 1900–1940, there were only 7 cases of flood hazards, affecting about 1955 people (Table 6.1). During the independent era, the expansion of corporate plantation and timber production resulted in increasing rates of deforestation, while urbanization began to rise, leading the country to be prone to flood-induced disasters. Having 3 cases of flood disasters from 1945 to 1965, the affected people by such disasters increased to about 3399 people.

In addition, the industrialization and agricultural expansion of local farming and corporate plantation and timber production occurred at greater scales during the New Order era of 1970–1999, continuing to the reformation era onward. Over-exploitation of ecological resources, such as massive land-use clearing, deforestation, and soaring urbanization, caused the increasing occurrence of flood hazards. Floods become a common problem, not only for many urban dwellers but also for rural people. Frequent flood hazards have occurred mainly in major cities as they are mostly located in coastal areas, such as the capital city of Jakarta, the city of Semarang, Surabaya, Medan, and Makasar. Table 6.2 shows the remarkable increase in flood hazards from the 1970s to the 2020s, impacting many fatalities, affecting people, and significantly damaging buildings and infrastructures.

The World Bank (2022) has highlighted that floods have been the major cause of disasters in Indonesia over the last two decades. According to their findings, nearly one-fourth of the country's population—particularly the poor and low-income households—faces a high risk of flooding. In one of Indonesia's poorest provinces, West Nusa Tenggara, for example, catastrophic flooding in the city of Bima in 2016 caused economic losses of about USD 65 million, severely affecting around two-thirds of its urban

Table 6.1 Long-term growth of Indonesia's economy, the incidence of flood hazards, and its impacts during the period of 1900–2023

Period	Indonesia's economy			Flood disaster impacts			
	Major production styles	Growth (%)	Urbanization (%)	Flood cases	Death	Affected people	Building damage
1. Dutch colonial era	Traditional farming, Colonial plantation, and slavery	1.7	2.7	7	NA	1955	NA
1900–1928		− 3.4	7.8				
1929–1940							
2. Independent Era	Nationalization of colonial plantation	1.3	15.9	3	NA	3923	NA
1945–1965							
3. New order era	Import-substitute industry, Green revolution, Corporate plantation	4.4	31.0	55	1965	31,777	39
1970–1990		3.0	41.9	106	366	14,505	3372
1990–1999							
4. Reformation era	Export-oriented manufacture, expansion of corporate plantation	3.5	49.8	3715	2864	32,495	402,127
2000–2010		2.0	56.7	10,163	1854	48,401	310,250
2010–2023							

[*Note*: There are significant differences in flood reportings before 1970s and after 1970s. The current reporting system since 2000 is based on data cards. Therefore, the frequency of floods prior to 1970s is obviously less than after 1970s. Systematic disaster data documentation started in 1970s]

population (The World Bank 2022). Urbanization exacerbates these challenges, as frequent flooding events can lead to widespread disasters in urban areas. This is often due to inadequate drainage systems unable to handle heavy rainfall and storms. Moreover, insufficient management of urban spaces, coupled with persistent urban poverty, tends to expand slum areas near urban rivers, further increasing many cities' vulnerability to flood hazards.

Moreover, urbanization, particularly the rapid development of high-rise buildings in coastal cities, has persistently depressed urban landscapes below sea level, thereby increasing their vulnerability to coastal flooding. Frequent urban flooding can escalate into disasters, causing severe damage to housing and the urban environment and often leading to serious public health crises. In such cases, the poor are disproportionately affected due to insufficient income, savings, and lack of insurance to cope with urban flood hazards. As population growth and urbanization continue, it is estimated that approximately 27% of Indonesia's population is highly prone to flood hazards, particularly those living in urban and coastal areas (Rentschler et al. 2022). The megacity of Jakarta serves as a striking example, having faced severe flood hazards in 2002, 2007, and 2013. These events caused significant economic losses, numerous fatalities, and widespread inundation. The 2002 flood, for instance, inundated approximately 331 km^2 of the city, resulting in 80 fatalities, 381 refugees, and economic losses of about USD 1.3 billion. Similarly, the 2007 flood caused economic losses of approximately USD 620 million, 79 fatalities, and displaced around 590,000 people (Budiyono 2005).

6.2.2 Socioeconomic Constraints in Promoting Insurance Program on Flood Disaster

6.2.2.1 Problems Associated with Provision of Ex Post Public Compensation to Flood Survivors

An insurance scheme for flood hazards is one of several options to manage the risks of disaster impacts by providing a legal right to compensation in the case of such disasters (Tesselaar et al. 2020). This compensation can help provide resources to increase the affected people's self-financing capacity to repair assets and build damaged homes (Patankar and Patwardhan 2016). Participation in insurance schemes enables individuals to reduce vulnerability through risk sharing among insurance policyholders, thereby reducing the socioeconomic impacts of disasters (Gianni et al. 2023). However, the recovery process of affected individuals from disaster impacts will be greatly influenced by the levels of household income and the availability of incentive-support program.

Households with adequate income can utilize insurance benefits to accelerate recovery (Platzer and Knodt 2023). Gurenko (2007) emphasizes further that the insurability of potential survivors of flood hazards, especially for those with low levels of income, depends upon the availability of government incentives to purchase insurance products. However, such assistance may be insufficient and ineffective measures to maintain the living standards of survivors, particularly the impoverished population. Similarly, Holzheu and Turner (2018) reveal that government assistance in post-disaster is often allocated primarily for emergency purposes and public facility repairs. Only a small portion of assistance designated for individuals is given to the survivors. People unable to obtain optimal benefits from the risk-sharing strategy may find themselves in worsening economic conditions. In such cases, survivors are often forced to immediately sell valuable assets to cover living expenses, pressuring them into debt traps and thus failing to overcome the impact of natural hazard-induced disasters. In addition, having internet access can influence risk awareness by providing a wider network and information related to disaster risks and benefits of purchasing disaster insurance (Holzheu and Turner 2018).

Meanwhile, having sufficient savings and accessible to loans can be alternative risk transfer strategies to be substituted for the ownership of disaster insurance schemes. The levels of household income largely influence the size of savings and payment to insurance premium. In this regard, because of having limited income flows, individuals often prefer to use savings as financial source to cover potential losses against potential disasters. Consequently, they tend to be less interested in participating in disaster insurance schemes (Holzheu and Turner 2018). However, a study by Holzheu and Turner (2018) emphasize that non-financial constraints can also lead to low rates of participation in disaster insurance. Such barriers include individuals lacking trust in insurance products and limited education and financial literacy about the benefits of disaster insurance (Eling et al. 2014). Some homeowners may also misunderstand that their insurance policies do not cover damage to their homes due to disasters. Additionally, insurance providers' service quality also contributes to enhancing individuals' interest in insurance products, including effective communication with customers, administrative procedures, and ease of purchase requirements (Holzheu and Turner 2018).

6.2.2.2 The Limit of Insurance Coverage to Outreach the Poor

Insurance coverage has limitations in helping the poor offset economic losses caused by natural hazard-induced disasters. As Holzheu and Turner (2018) point out, one key issue is that most insurance premiums are calculated based on the type and magnitude of risk rather than the payer's financial capacity. Additionally, high transaction costs can lead to increased premiums, making insurance unaffordable for lower-income communities. On the other hand, setting premiums based on actual risk levels has its advantages. It helps ensure the sustainability of insurance programs and raises awareness about disaster risks, potentially encouraging efforts to mitigate those risks (Box et al. 2016).

Despite these challenges, relevant stakeholders, including the government and insurance providers, can undertake various efforts as outlined by Holzheu and Turner (2018). Firstly, the government may increase participation in insurance by enacting regulations mandating participation in insurance programs. However, few countries have mandated disaster risk insurance. Most countries have only implemented mandatory insurance in social security programs, including health, employment, and motor vehicles. Compulsory vehicle insurance in developing countries is also commonly used as a gradual strategy to raise education to introduce the insurance products.

Meanwhile, mandatory insurance programs require other efforts to ensure their effectiveness, such as product design with premium options that can be adjusted based on individuals' capabilities. In this regard, government subsidies may be an option only when needed or when the inability to pay premiums becomes a major issue. The government may also establish specific building standards that are more disaster-resistant, require residential areas to be located far from disaster-prone areas, and encourage accurate data provision by home developers regarding its safety level to reduce potential losses due to disasters (Flood Insurance Work Group 2017).

Insurance providers can also promote risk mitigation by calculating costs or assessing premiums to encourage prevention and risk management (Paudel 2012). However, the challenge lies in cases where communities are less optimistic about the benefits of investing in risk mitigation through insurance. Therefore, innovation in insurance products can be a strategy to increase participation in disaster insurance. Holzheu and Turner (2018) mention that bundling product designs may automatically provide insurance to target community as an additional product option or integrated into their networks. An example is employment insurance which may include old-age benefits and workplace accident risk coverage. Designing agricultural insurance products combined with microcredit products or agricultural raw material distributors is also expected to increase access and farmers' participation. In addition, mortgage banks can play a role in bundling credit and home insurance products to raise participation in long-term insurance with lower transaction costs.

Furthermore, microinsurance schemes aimed at low-income societies and micro, small, and medium-sized enterprises, as well as designed to be easily accessible and employed for these groups, can be optimized to increase the participation of the poor and vulnerable in household disaster insurance (Ministry of Finance Indonesia 2018). Microinsurance provides relatively easier and more affordable premium schemes, though the coverage value is lower than standard insurance products. To reduce risk in the payment and claims process, stakeholders, including insurance providers, often use community group systems, which are considered more effective in this scheme Holzheu and Turner 2018). However, microinsurance programs also face challenges, including sustainability, especially if dependent on government policies or other stakeholders, and accurately targeting and classifying between subsidy recipients and those receiving full assistance if they cannot afford the insurance premium (Islam et al. 2022).

6.3 Methods

6.3.1 Study Area: Kampar, Riau, Indonesia

This research took place in Kampar Regency, Riau Province, as one of the areas that frequently experiences annual flooding. There were five villages used as survey locations, including Gunung Sahilan Village, Sungai Lipai Village, Sahilan Darussalam Village, and Kebun Durian Village. The four are in the Gunung Sahilan sub-district, an area prone to annual flooding.

6.3.2 Data Collection Methods: Quantitative and Qualitative

This study uses qualitative and quantitative data derived from primary and secondary data. Secondary data was obtained through a literature review, including the publication of previous research, books, reports, and statistical data from relevant and credible institutions. Meanwhile, primary data collection was carried out through in-depth interviews, Focus Group Discussions (FGD) with key informants and stakeholders, and surveys through household questionnaires.

The survey data include flood history, vulnerability, and socioeconomic indicators. 101 household respondents were selected by random sampling method in five villages as the survey locations during March 2023. A face-to-face survey was carried out by local enumerators experienced and trained to understand the questions posed by the research team and to reach the targeted respondents. At this stage, paper-based questionnaires were used to make notes quickly and prevent the risk of reaching respondents' locations with limited internet networks. In the next stage, the enumerator inputs the data online into the Google Form to fasten the data entry and to help the research team monitor the process. This survey consists of questions related to socioeconomic characteristics, living conditions, disaster experience, needs for disaster insurance, and their willingness to pay and response to disaster.

6.3.3 Data Analysis

The analysis examined respondents' data regarding their profile, living conditions, economic characteristics, disaster experiences, willingness to pay for disaster insurance, and disaster response. Participants were asked to report their information, i.e., age, education, job, and the number of people in their household, including themselves. Regarding the housing situation, the survey included queries regarding ownership status, house area, characteristics, and legality of the house. Respondents were also asked whether they considered the disaster risk before choosing the residence. Meanwhile, the economic situation variable encompasses income, expenses, alternative livelihood, insurance program participation, and disaster awareness. It aimed to capture various aspects of participants' financial circumstances and their engagement with insurance and disaster-related information.

The survey data analysis consists of two parts. The first part, which constitutes the main findings of this survey, involves simple descriptive analysis. This approach aims to present, in a straightforward manner, the changes observed in insurance demand-related variables after the disasters (Li and Liu 2023; Mumo and Watt 2019). Moreover, for the insurance willingness to pay (WTP), respondents were given open-ended questions to indicate the amount they would be willing to pay for an annual earthquake insurance scheme. The multiple answers to the questions were given, including limited coverage for the building (excluding furniture), with a coverage percentage of 50% of the building's value and a deductible of 5% of the damage incurred. It is worth noting that this question was only posed to those who reported needing an existing insurance policy. Analyzing these WTP responses seeks to provide valuable insights into the barriers hindering the proposed development of disaster insurance schemes.

6.4 General Result: Current State of Flood Risk Management in Kampar, Riau

6.4.1 Historical Flood Events and Impacts

Uncontrolled changes in land use and area cultivation may cause the flooding issues and related damages. Furthermore, rising rainfall intensities within cities and in the upstream catchments of rivers are also exacerbating the issue of flooding (Cummings et al. 2015). The river's overflow submerges the settlement areas in the floodplains and lower highlands as a result of these events.

The province's capital, Pekanbaru, has a distinct quality influenced by two major catchments: the upstream and downstream. The Malacca Strait forms the downstream boundary, and the hills form the upstream catchment boundary. Pekanbaru is a hub for both economic and governmental activity. About 160 km are found upstream of the Siak River in Pekanbaru. One of the largest rivers in the province of Riau is the Siak River. The Siak River flows from the west to the east, passing through Pekanbaru. More and more people with various economic pursuits live on both the left and right sides of the river basin. This river's primary use is for navigation, crucial to the area's social and economic growth. The communities that surround the river are engaged in trade and traditional fishing (Junaidi et al. 2018).

The flood happened beside the riverbanks, but significant rainfall intensities also impacted Pekanbaru, which also impacted the catchment areas (Teriyoko and Mudjiatko 2015). High tides also prevented floodwater from flowing downstream. Even with all of the preparations made in advance of the flood, such as the construction of flood mitigation facilities and water pumps, the real issues could not be avoided. The infrastructure and facilities were insufficient in relation to the demands in place (Junaidi et al. 2018).

In the flooded Pekanbaru area, the average water level height was recorded as + 4.60 m in October 1985, + 4.40 m in March 1988, + 3.80 m in November 1994, and + 3.80 m in December 2004 above sea level. The floods in Pekanbaru are progressively rising in water level while the flow velocities are falling. In Pekanbaru, inundations often last for 14 days. The unimpeded flood mitigation infrastructure, such as dikes, pumps, and water gates, is the reason behind the sensitive and vulnerable areas. Pumps and water gates are examples of incomplete flood control infrastructure, contributing to sensitive and vulnerable areas.

The floods in Pekanbaru, specifically in the public service areas and inhabited settlements, typically happen during periods of intense local rainfall and high seas. Over the past 10 years, Pekanbaru has seen an increase in the frequency of flooding occurrences brought on by heavy local rainfall (Junaidi et al. 2018). This increase in the development of infrastructure has occurred despite the potential consequences for the environment and the hydrological system. Flooding typically results in significant damages, which can be described as follows:

1. Houses and public infrastructure, including roads, schools, public health facilities, and wells, have been destroyed because of the flooding in the heavily populated areas.
2. These damages and destructions have disrupted the community of Pekanbaru's socioeconomic activities. Most of the people of Pekanbaru have taken refuge in the safer areas.
3. Many infections have been reported due to the flooding and the stench it has caused.

6.4.2 Existing Flood Preparedness and Response Measures (Qualitative)

Junaidi et al. (2018) emphasized the importance of adopting a comprehensive flood hazard mitigation management approach for pre-, during-, and post-flood situations in Indonesia. Pre-flood measures include preventive actions, such as land use management to restrict habitation in flood-prone areas and constructing dikes along rivers to protect nearby communities and infrastructure. Mitigation strategies aim to reduce the likelihood of permanent damage to property and loss of life, with examples including land acquisition and regular monitoring and inspections. Preparedness involves precautionary steps to enhance operational capacities and ensure effective responses during flooding, such as implementing warning systems and evacuation protocols. During and post-flood phases, actions focus on saving lives, minimizing property damage, and supporting recovery efforts, with emergency medical assistance serving as a critical response measure.

Planning strategies for flood mitigation in Pekanbaru should consider the following five factors: Economic aspects: Enhance financial compensation mechanisms and provide additional land to relocate businesses and settlements affected by flooding. Social aspects: Engage residents of flood-prone areas in public participation processes, as they are directly impacted by flooding. Additionally, integrate flood prevention strategies with poverty alleviation and social rehabilitation programs. Legal considerations: Ensure government regulations related to land status and compensation guidelines are legitimate and enforceable. It is also crucial to enforce laws addressing illegal activities, such as deforestation, in the catchment areas. Technical factors: Implement, manage, and maintain structural flood defense measures to achieve flood protection levels capable of withstanding events occurring once every 50 years.

6.4.3 Community Awareness and Perception of Flood Risks

Risk perception is frequently invoked in disaster literature to elucidate people's perceptions of and reactions to disaster risk (Hansson et al. 2020). Even though most risk analysis methodologies focus on objective measures, risk perception is currently regarded as an essential component of subjective risk analysis in the context of sustainable flood risk management (Kellens et al. 2011). The process by which every individual notices and internalizes risk, such as the mix of hazards, vulnerabilities, and exposure, can be collectively referred to as "risk perception" (Thistlethwaite et al. 2018).

Risk perception is studied using a variety of theories. Using behavioral choice theory and protective motive theory, Zinda et al. (2021) examined recent flooding experiences, current risk perceptions, and local adaptive responses in Troy, New York. Behavioral choice theory aims to correlate individual risk perceptions (magnitude and likelihood) with real risk estimations. Additionally, it draws attention to cognitive biases that lead individuals to behave in ways that deviate from their typical course of action when faced with calculated risks. Furthermore, the protection incentive hypothesis also considers the perceived efficacy of protective measures (Rogers 1997).

When people conduct a "coping appraisal" and determine that the adaptive strategies they have access to are (a) expected to be effective (response efficacy), (b) economical (sufficiently minimal response cost), and (c) practical to implement (self-efficacy), they will act defensively, according to Poussin et al. (2025). Threat appraisal, or risk perception, has a more substantial influence on protective behaviors than coping appraisal, according to Bubeck et al. (2012). Furthermore, to gain a deeper understanding of how communities manage risks, Shah et al. (2022) define the political ecology theory by examining the contextual factors (such as structural limitations, political pressures, and economic characteristics) that influence the vulnerability of various disadvantaged groups.

In light of the sociocultural context, understanding how society views the risk of flooding is crucial for determining exposure to and potential consequences of floods (Renn 2005). Knowing how society views a topic helps determine the most practical way to educate the public on the effects of flooding, promote preparedness, and increase public confidence in the government (Bradford et al. 2012). Since health emergencies are one of the most important flood risk, understanding how flooding affects the prevalence of diseases and mortality in the communities vulnerable to floods is another part of the social context. Individuals respond based on their perception of the health hazards they face, shaped by their educational, cultural, sociological, and economic backgrounds (Shah et al. 2022). Developing mitigation strategies requires understanding the social variables that influence how people perceive their health risks (Ferrer and Klein 2015; Shah et al. 2022).

The failure of costly, publicly supported structural approaches to mitigate risks has triggered a paradigm shift in a number of nations, according to Asmara et al. (2022). The fundamental idea

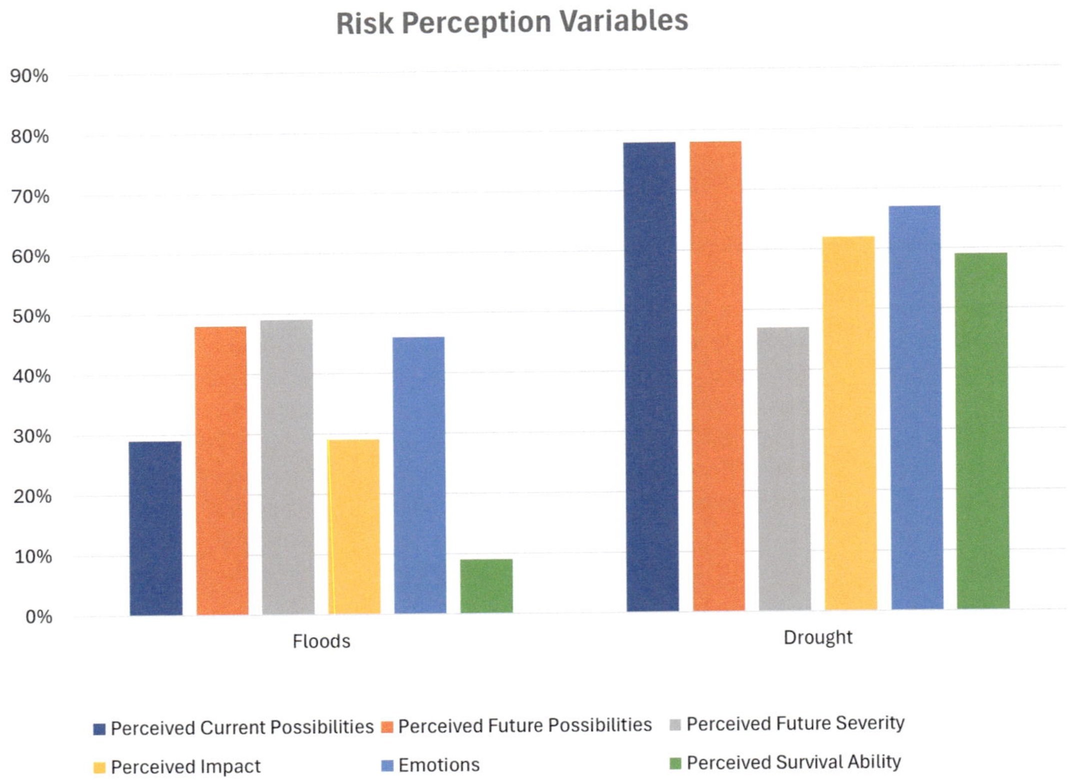

Fig. 6.4 Respondents' percentage of affirmative responses regarding their perceptions of the risks of flooding and drought

behind this paradigm shift is that multiple stakeholders must share responsibilities. The occurrence of floods and drought in Bandung has impacted the public's view of risk and pushed the city government to take many efforts to overcome these risks, according to the survey results shown in the above picture. Affected communities can evaluate the danger based on the survey results (see Fig. 6.4). On the other hand, compared to drought danger, the community's perception of flood risk is lower. Affected communities also support city government programs that are carried out because they demonstrate community approval of the initiative. Community faith in the program and the government and personal accountability to participate in mitigation measures are important driving forces behind this. Community satisfaction is lower than all measures (see Fig. 6.5). Nonetheless, none of the initiatives in place have been able to stop droughts or floods thus far.

There is a chance that Bandung City's affected communities will be participating in efforts to manage flooding and drought. This is demonstrated by their understanding of their duty to act and their conviction that community involvement is essential at every level of program development. This will lessen the desire to take action, despite a low perception of flood risk and a high public support of government programs contrary to the protective motive theory. On the other hand, this potential can be utilized to promote shared accountability while taking risks. Engaging the community can be a good method to share the costs of risk reduction measures themselves, enhance risk information, develop community resilience, and raise awareness about hazards. As a result, it is determined that Bandung City's

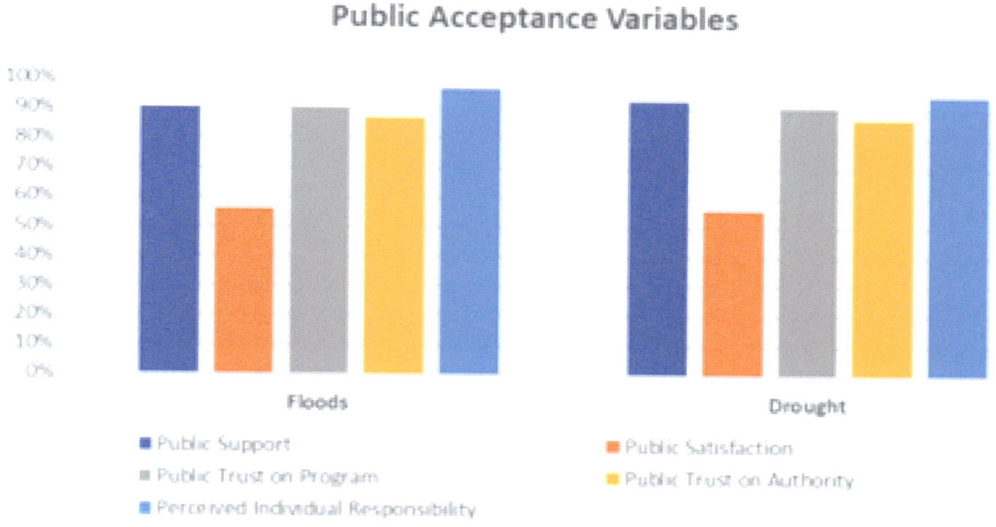

Fig. 6.5 Percentage of participants that selected affirmative responses on floods and drought

community-based disaster mitigation initiatives require improvement.

6.4.4 Institutional Framework for Flood Disaster Management

The flood solution is not just about making an area flood-free. So, it is not an exaggeration to say that the problem of flooding is the same as the problem of human life. The institutions that manage the occurrence of floods are the Water Resources Service and the Public Works Department through Balai Besar. The occurrence of local floods, tidal waves, and backwater is managed by the Department of Settlements and Spatial Planning and the City or Regency Government. Flood disaster management involves many parties. These parties are the government, DPR/DPRD, society, universities, associations, private parties, and other institutions. The stakeholders involved are presented in Table 6.2:

Disaster management institutions in Indonesia have existed since 1966, when the government established a body to handle natural disaster management issues, which was renewed in 1979. In 1990, Presidential Decree No. Natural hazard-induced disasters and disasters that arise due to human activities. The scope of duties was then expanded again through Presidential Decree No. 106 of 1999, which included survivors of social conflict in an area. Then, in 2001, Presidential Decree No. 3 concerning the National Coordinating Agency for Disaster Management and Refugee Management (BAKORNAS PBP). With the issuance of Law No. 24/2007 concerning Disaster Management, a reasonably comprehensive law in managing risks from natural hazards, non-natural hazards, and social conflict, the institution responsible for handling disaster management at the central level is the National Disaster Management Agency. And in the regions is the Regional Disaster Management Agency. The legal basis for forming this Agency is paragraph 4 of the Preamble to the 1945 Constitution. Based on Law 24 of 2007, the Presidential Regulation of the Republic of Indonesia was stipulated no. 8 of 2008 concerning the National Disaster Management Agency. The National Disaster Management Agency (BNPB) is a non-departmental government agency headed by a head and directly responsible to the President (PUPR 2017).

Table 6.2 Stakeholders involved in flood disaster management (*Source* PUPR 2017)

1. Private and institutional other	DPR, DPRD Government: Department 2:	1. Provincial Government District/City Government
– Investors – Equipment industry – Contractor – Supplier consultant – Volunteer – Company Private, ETC 2. Public – Urban – Rural – NGO, etc. 3. College public and private 4. Research institute 5. Associations: IAI, PII, HATHI, HAGI, HATTI, Gapensi, Gapeknas, Inkindo, etc.	– Agriculture – Health – Social – Forestry – Maritime affairs and fisheries – Finance – Information/information and communion – Labor and transmigration – Environment – Tourist – Transportation and telecommunications – Energy resources and mineral – Public works – Judiciary – Ministry of environment office – BMG prosecutor's office post	Provincial Bappeda, District/City Departments Prop, District/City: 2. PU Irrigation, Bina Clan Health, Social, Agriculture 3. Creation of works, mining 4. Hospital 5. Sub-district head, village head, and the device 6. Tirta services 7. Infrastructure projects 8. PLN TNI Polri 9. SAR team 10. Indonesian red cross firefighter

6.5 Result and Discussion on the Feasibility Assessment of Flood Insurance Scheme

6.5.1 Community Willingness to Participate in a Flood Insurance Scheme

The ownership of a house is a significant determinant for individuals seeking to obtain insurance coverage for their homes to mitigate the potential consequences of natural or man-made hazards. Based on a survey carried out in Kampar, Riau, it was found that a significant majority of respondents, specifically 87.13%, expressed the view that the acquisition of disaster insurance is necessary. Of the individuals who voiced the necessity for disaster insurance, a majority of 76.24% indicated their willingness to participate in a program should one be available. Remarkably, a mere 67.33% of the overall participants who indicated an interest in enrolling in the disaster insurance program showed a readiness to bear the associated insurance premium.

Table 6.3 presents the data indicating that individuals who own homeownership are more inclined to acknowledge the necessity of insurance, demonstrate a desire to participate in the program, and express a readiness to make financial contributions toward the insurance, with percentages of 82.95%, 84.42%, and 83.82%, correspondingly.

Table 6.3 Prevalence of respondents to participate in disaster insurance program

Type of house ownership	Disaster insurance needs (%)	Willing to join disaster insurance program (%)	Willing to pay disaster insurance premium (%)
Official house/free house/family house	12.50	12.99	13.24
Owned house	82.95	84.42	83.82
Rent house	4.55	2.60	2.94

The proximity to a natural hazard significantly impacted the participants' perceived need for insurance, their willingness to engage in the program, and their preparedness to spend financial resources toward covering the cost of insurance. All participants who expressed a requirement for disaster insurance had experienced a flood incident within a period of 1–2 years prior to the survey. Similarly, within the group of respondents who indicated their desire to engage in the disaster insurance initiative, the complete sample, or 100.0%, reported experiencing a flood incident within the period of 1–2 years preceding the study. In a similar vein, the entirety of those who responded (100%) in the insurance program survey who expressed their readiness to pay the insurance premium had experienced a flood incident within a period spanning from one to several years prior (Table 6.4).

Table 6.4 Prevalence of respondents with disaster experience to participate in disaster insurance program

Disaster experience	Disaster insurance needs (%)	Willing to join disaster insurance program (%)	Willing to pay disaster insurance premium (%)
1–2 years ago	100.0	100	100

Insurance prices might pose a significant barrier to participation in insurance (Kalfin et al. 2022; Mccord et al. 2005). The survey findings offer support for this notion, as respondents shown a willingness to allocate supplementary funds ranging from IDR 5000 (< USD1) to IDR 200,000 (USD13.33)[2] on a monthly, quarterly, or annual basis. After considering temporal variables, the results of the survey suggest that the average insurance premium (monthly adjusted) is projected to be IDR 8811 (< USD1) per month, while the median insurance premium is IDR 10,000 (< USD1) per month. This discovery affirms that the willingness to pay for participation in a disaster insurance scheme is relatively low in Indonesia. Therefore, it is likely that microinsurance for disaster financing mitigation is suitable for implementation in Indonesia. However, it is important to note that this program can be effectively implemented nationally through the government of Indonesia's intervention in the universal disaster insurance program.

Furthermore, upon analyzing the participants' willingness to allocate financial resources based on their income and expenditures, it is approximated that the corresponding ratio stands at approximately 0.3–0.4% for each category. This demonstrates the extent to which individuals allocate their financial resources to acquire insurance coverage (Tables 6.5, 6.6).

Table 6.5 Prevalence between disaster insurance premium and period of premium payment *Source* Authors' calculation

Insurance premium	Yearly (%)	Every 3 months (%)	Monthly (%)
5000	0.0	0.0	4.4
10,000	5.9	16.2	50.0
15,000	0.0	4.4	0.0
20,000	1.5	1.5	7.4
30,000	0.0	1.5	0.0
50,000	1.5	4.4	0.0
200,000	1.5	0.0	0.0

Table 6.6 Estimated willingness to pay, ratio of premium to expenditure and income *Source* Authors' calculation

	Ratio
Ratio premium to expenditure	0.4%
Ratio premium to income	0.3%
Estimated average premium	IDR 8811/month
Estimated median premium	IDR 10,000/month

In accordance with the respondents' inclination to allocate funds for disaster insurance, they also considered the constraints associated with insurance coverage. In particular, a significant majority of participants, amounting to 73%, who indicated a readiness to pay insurance premiums, expected to receive coverage amounting to less than IDR 10 million (< USD 666.67) for each occurrence of a flood event. In addition, 7% of the participants expressed their anticipation of obtaining insurance coverage within the range of IDR 10 million to IDR 20 million (equivalent to USD 666.67–1333.33). The survey conducted in the Kampar District yielded results indicating that the

[2] Exchange rate estimation is 15,000 IDR/USD.

respondents exhibited a comparatively lower level of readiness to pay compared to market-based insurance businesses and industrialized nations (Table 6.7).

Table 6.7 Prevalence between disaster insurance premium and insurance coverage *Source* Authors' calculation

Disaster insurance premium	Insurance coverage	
	Between Rp. 10–20 million (%)	Less than Rp. 10 million (%)
5000	0.0	4.4
10,000	1.5	70.6
15,000	0.0	4.4
20,000	4.4	5.9
30,000	0.0	1.5
50,000	1.5	4.4
200,000	0.0	1.5

Based on the survey conducted in Riau in 2023, the results suggest that the development of an effective insurance program in Indonesia necessitates consideration of the income and expenditure levels of the populace. According to Alam et al. (2020), Kalfin et al. (2022), Kashyap et al. (2006), and McCord et al. (2005), microinsurance and national insurance programs are viable choices for Indonesia, a developing nation, and can function as substitutes for disaster risk funding. These programs have the potential to effectively respond to the unique requirements and economic limitations of the populace in terms of insurance protection against flood events.

6.5.2 Affordability and Accessibility of Insurance Products

Implementing flood insurance schemes is essential for mitigating the adverse effects of such disasters. In terms of affordability and accessibility of flood insurance for rural households, rural communities' socioeconomic could be associated with insurance scheme uptake. It can be seen from the graph below that the dominant income level from the respondents is between Rp 2,000,000 and Rp 3,000,000 counted for 41% respondents while only 1% of the respondents with income less than Rp 1,000,000. There are 12% of respondents with more than Rp 5,000,000 monthly income, and 10% calculated in the income range between Rp 4,000,001–Rp 5,000,000. Limited income levels and financial resources in these areas pose a significant barrier to the affordability of insurance premiums Fig 6.6.

A direct correlation between household income and insurance uptake highlights the challenges lower-income groups face in accessing insurance products. Furthermore, the predominant economic activities in rural areas, often centered on agriculture and fisheries, influence the perception and accessibility of insurance. The table below provides insights into how economic status may affect the decision-making process regarding risk disaster insurance. For the common natural disaster, people are aware of the possibility of a common natural disaster in their area, which is flood. However, only 4.91% of respondents recognize more potential of natural disasters such as bushfire and landslide in their area (Table 6.8).

Geographical factors also significantly affect the implementation and success of insurance schemes. Areas more prone to flooding are often faced with higher insurance premiums, making insurance less accessible to poorer households. It can be seen from the table above that 95% of respondents feel insecure about the possibility of flooding in their area. The insecurity level is almost even between respondents who felt worried about the possibility of a flood (48.51%) and respondents who felt less concerned with the flood in their area (51.49%). This condition may conclude the reason for 87.13% of respondent who stated that risk insurance is needed while 12.87% disagree with the need for disaster risk insurance.

The capacity of local institutions and the level of community engagement are crucial in determining the effectiveness of insurance schemes. The table below shows that 28.7% of respondents rely on social assistance to support their basic needs (food and water, house, cloth) while evacuating to safer places in the first 6 months of disaster. The length of this condition varies

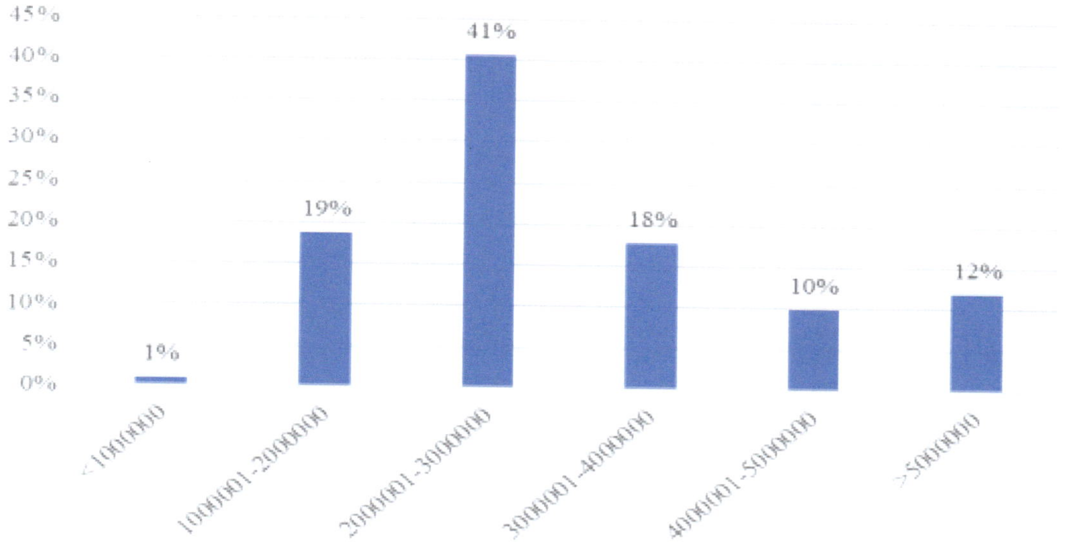

Fig. 6.6 The respondent's level of income

Table 6.8 Risk awareness on potential disaster based on the level of income *Source* Authors' calculation

Total monthly income	Risk awareness on potential disaster		Level of insecurity		Risk insurance needs	
	Flood	Flood, bushfire, and landslide	Worry	Less worry	Yes	No
< 1,000,000	1	0	1	0	1	
1,000,001–2,000,000	17	2	9	10	18	1
2,000,001–3,000,000	39	2	18	23	34	7
3,000,001–4,000,000	17	1	9	9	14	4
4,000,001–5,000,000	10	0	5	5	10	0
> 5,000,000	12	0	7	5	11	1
Sub total	96	5	49	52	88	13
Percentage (%)	95	5	48.51	51.49	87.13	12.87

among respondents. In total, 51.48% of respondents mentioned asking for help and assistance for their basic needs. It may imply the role of local governments in promoting and subsidizing insurance to enhance the accessibility of support and assistance for the affected household. Moreover, community education and awareness programs may influence perceptions of insurance benefits and risks, especially in rural communities (Table 6.9).

While the affordability and accessibility of risk disaster insurance are challenging for rural communities, there are opportunities for improving the market of flood insurance schemes. The involvement of local stakeholders in the design and implementation of insurance products can ensure that they are tailored to the specific needs and circumstances of rural

Table 6.9 The response of respondent during the first 6 months of disaster *Source* Authors' calculation

Action*	Frequency	Percent
1	14	13.9
2	3	3.0
3	17	16.8
5	1	1.0
6	1	1.0
7	2	2.0
8	1	1.0
1, 2	8	7.9
1, 3	29	28.7
1, 5	1	1.0
1, 7	6	5.9
2, 3	1	1.0
2, 5	1	1.0
3, 6	1	1.0
3, 7	2	2.0
3, 8	1	1.0
1, 2, 3	2	2.0
1, 2, 4	1	1.0
1, 3, 5	1	1.0
1, 3, 6	4	4.0
1, 3, 7	2	2.0
3, 7, 8	1	1.0
1, 3, 7, 8	1	1.0
Total	101	100

**Notes*: The types of action or responses among respondent are coded below:

1. Evacuate to safer place
2. Take out saving
3. Looking for help/assistance (basic needs)
4. Selling asset
5. Selling livestock
6. Borrow money
7. Reduce secondary (and tertiary) expense
8. Asking for remittance from their family outside the affected area

communities, including the premiums and benefits. Government interventions, such as subsidies or low-cost insurance options, can improve affordability and accessibility by exploring innovative models and strategies that can overcome the barriers to insurance accessibility and affordability in rural and economically disadvantaged regions.

6.5.3 The Capacity of Insurance Institution for Natural Disaster

Indonesia has been identified as a nation with a high susceptibility to natural hazard-induced disasters, and the occurrence of floods is a prevalent phenomenon inside the country's boundaries (The World Bank 2021). Figure 6.7 presents the number of natural hazard-induced disasters in Indonesia throughout 2023. The Indonesian government recognized the necessity of identifying alternate methods to fund disaster response efforts. In recent years, the Ministry of Finance has established a comprehensive strategy known as the National Disaster Risk Finance and Insurance (DRFI) Strategy (The World Bank 2021).

A recent study by Cobian et al. (2022) examined the level of interest and demand among households in Jakarta, Indonesia, for a theoretical flood insurance product based on an index. According to Cobian et al. (2022), the research demonstrates that the utilization of this specific good has the potential to safeguard household income and consumption from flood shocks external to the household entity. The demand for flood insurance should follow household expectations regarding flood events. The increasing demand should align with the expectation or experience of flood events. Figure 6.8 reveals the phenomena of flood events during the period. The flood events fluctuated from 2015 to 2023, which may influence the willingness of households to buy insurance.

On the other hand, an alternative article posits that relying solely on financializing risk is not a comprehensive solution to address Indonesia's flood challenges. Instead, it suggests that the government should allocate resources toward developing forecasting technology, research endeavors, and human capital (Colven 2023). These investments would enhance the institutional capabilities of regional agencies, enabling them to effectively plan for and mitigate the impact of floods. Following the Colven (2023) argument, the increasing institutional capabilities should be followed by the proper financial instruments. Realizing the limit of government capacities for

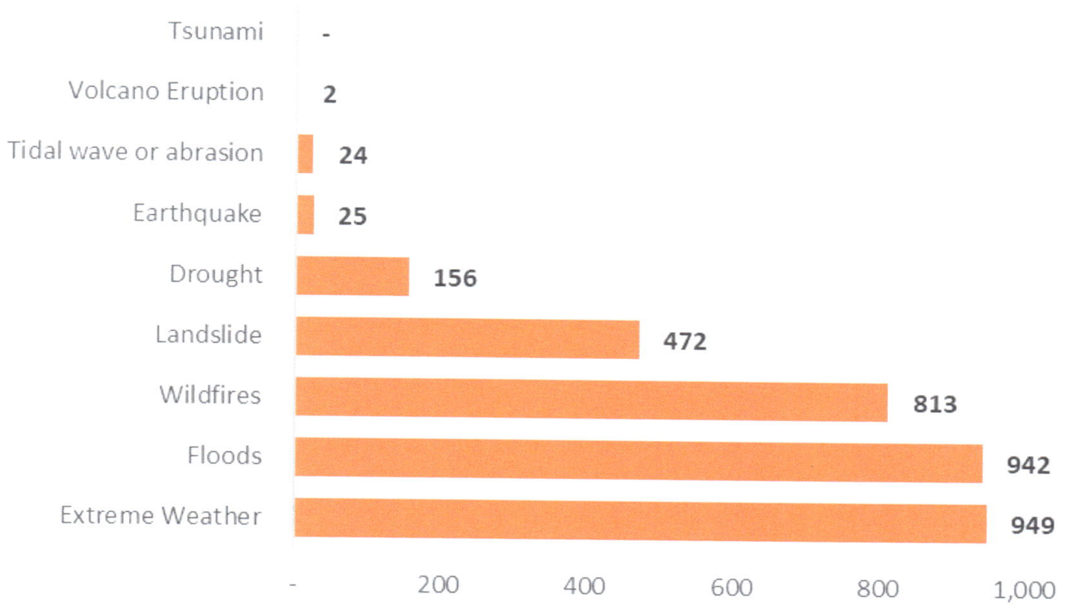

Fig. 6.7 Number natural hazard-induced disasters in Indonesia, 1 Jan–15 Nov 2023 (*Source* Annur 2023)

preventive investment, the risk transfer mechanism of natural hazard events such as floods, it should be done immediately. This is related to the speed of sea level increasing, extreme weather, and other disaster climates related to hydrometeorology events.

The complexity of risk transfer through insurance in developing nations is heightened by their limited financial capacity to effectively mitigate the associated risks. In relation to our research findings, the limited financial resources available for covering insurance premiums consistently provide a challenge in developing nations. Without an established insurance culture or and familial or governmental assistance, individuals affected by disasters may experience a deepening of poverty. This can occur if survivors resort to high-interest loans, sell off assets and livestock, and engage in low-yield farming practices as a result of the adverse impacts of extreme occurrences Varangis et al. 2003). According to Mechler et al. (2006), implementing microinsurance with restricted coverage and decreased transaction costs holds promise as a viable approach for facilitating risk transfer within developing or low-income markets. Cohen and McCord (2003) have identified four institutional models for implementing microinsurance. These models include the community-based model, full-service model, provider model, and partner-agent model. Regarding capacity, certain companies offer earthquake microinsurance products characterized by affordable insurance premiums and restricted coverage (see Table 6.10). In a broader context, this condition has the potential to apply to flood situations but with certain modifications and stipulations.

According to Mccord et al. (2005), to enhance marketability, the emergence of microinsurance had to be accompanied by cost-minimization strategies. Mccord et al. (2005) emphasized the significance of employing competent staff and agents in order to effectively engage with the low-income market. Cultural barriers often arise as a result of a prevailing lack of confidence toward insurance businesses, including microinsurance entities. Training staff and agents is crucial for the success of microinsurance enterprises involved in customer engagement. According to Mccord et al. (2005), there is a potential for training initiatives in Indonesia that could be facilitated by the Institute of Risk Management. This institute is a merger

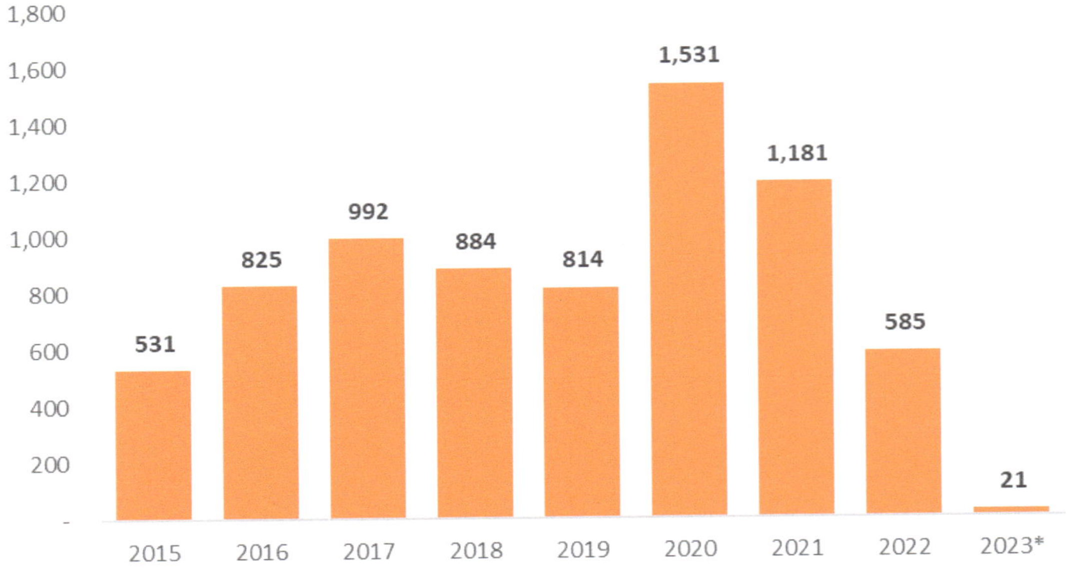

Fig. 6.8 Number of flood events in Indonesia, 2015–2023* [*Source* Annur (2023) * February 20, 2023]

of the Akademi Assuransi Indonesia (AAI) and Lembaga Pendidikan Assuransi (LPAI), and it offers insurance training programs to the industry, including Sharia-based microinsurance. Nevertheless, microinsurance businesses are limited in providing coverage to a small client to maintain reasonable levels of risk.

The potential diversification of company schemes given for scaling-up to a nationwide level with uncorrelated risk is a consideration. However, it is important to note that this can additionally give rise to a reduction in institutional familiarity and trust. Despite this potential disadvantage, it is possible that such diversification could contribute to success and expense reduction (Mccord et al. 2005). The expansion of operations on a national scale has the potential to amplify systemic risks, such as floods and other calamities. Consequently, private insurance providers may exhibit reluctance in offering coverage for these risks due to the potential emergence of moral hazards and adverse selection (Kunreuther 1998). Increased hazards occur when insured individuals modify their behavior, while adverse selection arises when confronted with heightened risks. According to Mccord et al. (2005), individuals only purchase insurance when they perceive a higher level of danger. The presence of asymmetric information hinders insurance companies' ability to accurately assess the extent of risk transfer from insurance buyers.

The government was compelled to implement the program because private insurers may not consistently be equipped to assume the associated risks. Several nations, including Japan, France, the United States, Norway, and New Zealand, have implemented government-regulated national insurance programs for natural hazard-induced disasters. These programs offer two types of schemes: mandatory participation, where individuals are required to enroll, and voluntary participation, where individuals can choose to enroll. The insurance coverage provided by these programs can be either limited to a specific hazard or comprehensive, encompassing a wide range of potential risks (Mccord et al. 2005).

Table 6.10 Earthquake microinsurance in Indonesia (*Source* Tri Pakarta, 2019; Asuransi Astra, 2019; ACA Asuransi, 2019 in Kusumawardhani et al. 2019)

Company	Product	Annual premium (IDR)	Coverage (IDR)	Type of dependents
Tri Pakarta	Micro "Si Bijak" insurance	50,000	500,000	Loss of livelihood
	Micro insurance stops earthquake, tsunami businesses	40,000	2,500,000	
	Cosmic micro insurance stop business	50,000	3,000,000	
Astra insurance	Micro insurance	30,000	–	– Loss of livelihood
				– Damage to buildings
ACA insurance	Earthquake fund micro insurance	20,000	2,000,000	

6.5.4 Discuss the Potential Benefits and Drawbacks of the Scheme: Feasibility

The findings provide valuable insights into the behavior of individuals vulnerable to disaster risks, particularly how they periodically react to potential disaster events. People adapt to these risks and challenges by taking measures to mitigate the impacts. While a few choose to relocate, the majority remain in their current locations. This may explain why people who have faced disasters over generations continue to choose not to move their settlements to safer environments.

Furthermore, the research highlights the feasibility of implementing risk microinsurance schemes based on the willingness of most homeowners to pay. Despite the low suggested insurance premium of IDR 10,000 per month, the willingness to pay for such coverage is a key finding of the study. A significant proportion of homeowners—more than 80%—expressed a willingness to pay for an insurance premium, reflecting an encouraging shift toward a culture of disaster risk awareness. However, the minimum insurance coverage amount of IDR 10 million is at least 1000 times greater than the suggested monthly premium, indicating a significant gap that needs to be addressed.

6.5.5 Discuss the Proposed Solutions and Strategies: For Example Importance of Community Engagement and Education

Despite a high rate of willingness to pay for disaster insurance (more than 80%), the community needs to understand risk financing in a better way. One desired outcome for education and awareness is to drive a lesser demand for financial support from the government in the event of a disaster. Before turning to the government, either central or local, that has tight fiscal space, the one should try to better equip themselves with various community programs and insurance schemes. Assistance from the diaspora residing in other cities or countries is also possible a financial source.

6.6 Conclusion

Based on the qualitative and quantitative findings, we can conclude that the community nevertheless needs the existence of a disaster insurance. People are willing to pay a certain amount in exchange for insurance coverage for unforeseen disaster events. Despite their low-income status, house ownership has motivated people's willingness to pay for insurance premiums. In addition, by putting in place insurance to cover losses from disaster events and some disaster risk safeguard mechanisms, people could stay and continue their lives in their original places.

This research contributes to a better understanding of disaster risk reduction strategies by revealing a willingness to pay from directly affected people. The findings serve as evidence that can support policy-making process. Furthermore, empirical evidence from this study has enriched academics in related fields including disaster management, insurance, and behavior economics, among others. Developing countries could learn from this case study to create assessments of similar disaster insurance schemes.

The study covered only the customers' point of view; thus, it needs a follow-up to discover the business appetite of the insurance companies. Findings have revealed the willingness to pay findings from the customer side, the desired amount of insurance premium, and the insurance coverage amount. Afterward, the firms' owners and managers could utilize that information and consider whether to introduce a particular risk insurance product or not. To create a sustainable insurance product, business entities must ensure the viability of such a scheme. Future research should incorporate the possibility of public–private partnerships in building an affordable yet sustaining insurance scheme. Furthermore, based on a calamity assessment, whether a local or regional insurance scheme is feasible to be conducted instead of a nationwide coverage area.

References

Alam ASAF, Begum H, Masud MM, Al-Amin AQ, Filho WL (2020) Agriculture insurance for disaster risk reduction: a case study of Malaysia. Int J Dis Risk Reduct. https://doi.org/10.1016/j.ijdrr.2020.101626

Aerts JCJH, Botzen WJ, Clarke KC, Cutter SL, Hall JW, Merz B, Michel-Kerjan E, Mysiak J, Surminski S, Kunreuther H (2018) Integrating human behaviour dynamics into flood disaster risk assessment. Nat Clim Change 8(3):193–199. https://doi.org/10.1038/s41558-018-0085-1

American Academy of Actuaries Flood Insurance Work Group (2017) The national flood insurance program: challenges and solutions. American Academy of Actuaries. https://www.actuary.org/wpcontent/uploads/2020/09/flood9.20_0.pdf

Angelakis AN, Capodaglio AG, Valipour M, Krasilnikoff J, Ahmed AT, Mandi L, Tzanakakis VA, Baba A, Kumar R, Zheng X, Min Z, Han M, Turay B, Bilgiç E, Dercas N (2023) Evolution of floods: from ancient times to the present times (ca 7600 BC to the present) and the future. Land. https://doi.org/10.3390/land12061211

Annur CM (2023) BNPB: Tren Banjir di Indonesia Cenderung Menurun dalam Tiga Tahun Terakhir. Katadata.Co.Id. https://databoks.katadata.co.id/datapublish/2023/02/20/bnpb-tren-banjir-di-indonesia-cenderung-menurun-dalam-tiga-tahun-terakhir

Asmara LY, Rianawati E (2022) Public risk perception and public acceptance of the existing flood and drought mitigation measure in Bandung city. IOP Conf Ser: Earth Environ Sci 986(1). https://doi.org/10.1088/1755-1315/986/1/012044

Booth A (1998) The Indonesian economy in the nineteenth and twentieth centuries. Palgrave Macmillan, New York. https://doi.org/10.1057/9780333994962

Box P, Bird D, Haynes K, King D (2016) Shared responsibility and social vulnerability in the 2011 Brisbane flood. Nat Hazards 81(3):1549–1568. https://doi.org/10.1007/s11069-016-2145-z

Bradford RA, O'Sullivan JJ, Van Der Craats IM, Krywkow J, Rotko P, Aaltonen J, Bonaiuto M, De Dominicis S, Waylen K, Schelfaut K (2012) Risk perception - Issues for flood management in Europe. Nat Hazards Earth Sys Sci 12(7):2299–2309. https://doi.org/10.5194/nhess-122299-2012

Bubeck P, Botzen WJW, Suu LTT, Aerts JCJH (2012) Do flood risk perceptions provide useful insights for flood risk management? Findings from central Vietnam. J Flood Risk Manag 5(4):295–302. https://doi.org/10.1111/J.1753-318X.2012.01151.X

Budiyono Y (2005) Flood risk modeling in Jakarta: development and usefulness in a time of climate change, vol 45. PhD thesis, Vrije Universiteit Amsterdam

Champonnois V, Erdlenbruch K (2021) Willingness of households to reduce flood risk in southern France. J Flood Risk Manag. https://doi.org/10.1111/JFR3.12696

Cobian J, Resosudarmo BP, Halimatussadiah A, Olivia S (2022) Demand for index-based flood insurance in Jakarta. Indonesia working papers in trade and development

Cohen M, McCord MJ (2003) Financial risk management tools for the poor. In: Microinsurance centre briefing note #6

Colven E (2023) Financialising risk is no panacea for Indonesia's flood woes. East Asia Forum. https://www.eastasiaforum.org/2023/05/06/financialising-risk-is-no-panacea-for-indonesias-flood-woes/

Cummings TF (2015) Assessment of nitrate export in agricultural sub-catchments of the Grand River watershed: an isotope approach. Master's thesis, University of Waterloo

Eling M, Pradhan S, Schmit JT (2014) The determinants of microinsurance demand. Geneva Pap Risk Insur Issues Pract 39(2):224–263. https://doi.org/10.1057/gpp.2014.5

Ferrer RA, Klein WM (2015) Risk perceptions and health behavior. Curr Opin Psychol 5:85–89

FWI/GFW (2002) The state of the forest: Forest Watch Indonesia, Bogor, Indonesia, and Global Forest Watch, Washington DC. http://pdf.wri.org/indoforest_full.pdf

Ghapar AA, Yussof S, Bakar AA (2018) Internet of Things (IoT) architecture for flood data management. Int J Fut Gener Commun Netw. https://doi.org/10.14257/ijfgcn.2018.11.1.06

Gianni E, Tyrologou P, Couto N, Correia V, Brondi S, Panagiotaras D, Koukouzas N (2023) Natural hazard insurance: dissemination strategies using geological knowledge. In: Environment systems and decisions. Springer, London. https://doi.org/10.1007/s10669-023-09936-9

Glago FJ (2019) Household disaster awareness and preparedness: a case study of flood hazards in Asamankese in the West Akim Municipality of Ghana. Jamba J Dis Risk Stud. https://doi.org/10.4102/JAMBA.V11I1.789

Glago FJ (2021) Flood disaster hazards; causes, impacts and management: a state-of-the-art review. In: Farsangi EN (ed). IntechOpen, New York. https://doi.org/10.5772/intechopen.95048

Gomez-Cunya LA, Tilt J, Tullos D, Babbar-Sebens M (2022) Perceived risk and preferences of response and recovery actions of individuals living in a floodplain community. Int J Dis Risk Reduct 67:102645. https://doi.org/10.1016/J.IJDRR.2021.102645

Gurenko E (2007) Climate change and insurance: disaster risk financing in developing countries. Earthscan, London

Hansson S, Orru K, Siibak A, Bäck A, Krüger M, Gabel F, Morsut C (2020) Communication-related vulnerability to disasters: a heuristic framework. Int J Disaster Risk Reduct 51:101931

Holzheu T, Turner G (2018) The natural catastrophe protection gap: measurement, root causes and ways of addressing underinsurance for extreme events†. Geneva Pap Risk Insur Issues Pract 43(1):37–71. https://doi.org/10.1057/s41288-017-0075-y

IPCC (2021) Climate change 2021: the physical science basis. In: Contribution of working group I to the sixth assessment report of the intergovernmental panel on climate change. Cambridge University Press. https://doi.org/10.1017/9781009157896

Islam MM, Matsushita S, Noguchi R, Ahamed T (2022) A damage-based crop insurance system for flash flooding: a satellite remote sensing and econometric approach. Asia Pac J Region Sci 6(1):47–89. https://doi.org/10.1007/s41685-021-00220-9

Jevrejeva S, Jackson LP, Grinsted A, Lincke D, Marzeion B (2018) Flood damage costs under the sea level rise with warming of 1.5 °C and 2 °C. Environ Res Lett. https://doi.org/10.1088/1748-9326/aacc76

Jongman B, Kreibich H, Apel H, Barredo JI, Bates PD, Feyen L, Gericke A, Neal J, Aerts JCJH, Ward PJ (2012) Comparative flood damage model assessment: towards a European approach. Nat Hazards Earth Syst Sci. https://doi.org/10.5194/nhess-12-3733-2012

Junaidi A, Daoed D (2018) Future flood management strategies in Indonesia. In: MATEC web of conferences vol 229, pp 1–7. https://doi.org/10.1051/matecconf/201822901014

Kalfin S, Supian S, Mamat M (2022) Insurance premium determination model and innovation for economic recovery due to natural disasters in Indonesia. Computation. https://doi.org/10.3390/computation10100174

Kashyap A, Anthony M, Krech R (2006) Microinsurance: demand and market prospect: India. Security, New Delhi

Kausarian H, Masdriyanto S, Suryadi A, Zainuddin Lubis M (2023) Geomorphology and geology studies using digital elevation model (DEM) data in the watershed area of Kampar Regency, Riau Province. J Geosci Eng Environ Technol 8(3):2023. https://doi.org/10.25299/jgeet.2023.8.3.14454

Kellens W, Zaalberg R, Neutens T, Vanneuville W, De Maeyer P (2011) An analysis of the public perception of flood risk on the Belgian coast. Risk Anal 31(7):1055–1068. https://doi.org/10.1111/j.1539-6924.2010.01571.x

Kron W (2005) Flood risk= hazard• values• vulnerability. Water Int 30(1):58–68

Kunreuther H (1998) Insurability conditions and the supply of coverage. In: Kunreuther H, Sr R, Richard J (eds) Paying the price: the status and role of insurance against natural disasters in the United States. The Joseph Henry Press, New York

Kusumawardhani D, Sagala S, Hafiz I, Ramadhani A (2019) Insurance for low-income families as an earthquake risk financing instrument. In: RDI working paper series, vol 4

Li Q, Liu W (2023) Impact of government risk communication on residents' decisions to adopt earthquake insurance: evidence from a field survey in China. Int J Dis Risk Reduct. https://doi.org/10.1016/j.ijdrr.2023.103695

MacKinnon K (1997) The ecological foundations of biodiversity protection. In: Last stand. Oxford University Press, Oxford. https://doi.org/10.1093/oso/9780195095548.003.0007

Marfai MA, Sekaranom AB, Ward P (2015) Community responses and adaptation strategies toward flood hazard in Jakarta, Indonesia. Nat Hazards 75(2):1127–1144. https://doi.org/10.1007/S11069-014-1365-3/FIGURES/8

Mccord MJ, Microinsurance T, Gtz GR, Mcguinness L, Opportunities M (2005) Microinsurance demand & market prospects Indonesia. Methodology 2005:1

Mechler R, Linnerooth-Bayer J, Peppiatt D (2006) Microinsurance for natural disaster risks in developing countries. In: A ProVention/IIASA study

Mendes-Da-Silva W, Lucas EC, de Carvalho JVF (2021) Flood insurance: the propensity and attitudes of informed people with disabilities towards risk. J Environ Manag 294:113032. https://doi.org/10.1016/J.JENVMAN.2021.113032

Ministry of Finance Indonesia (2018) Strategi Pembiayaan dan Asuransi Risiko Bencana. In: Kementerian Keuangan Republik Indonesia

Mumo R, Watt R (2019) Residential insurance market responses after earthquake: a survey of Christchurch dwellers. Int J Dis Risk Reduct. https://doi.org/10.1016/j.ijdrr.2019.101166

Mwape YP (2009) An impact of floods on the socio-economic livelihoods of people: a case study of Sikaunzwe Community in Kazungula District of Zambia. A Mini Dissertation for the Award of Masters Degree in Disaster Risk Management, Disaster Risk Management Training and Education Centre for Africa (DIMTEC), Faculty of Natural and Agricultural Sciences, University of the Free State

O'Hara PA (2023) Principles of institutional and evolutionary political economy: applied to current world problems. Springer, Singapore. https://doi.org/10.1007/978-981-19-4158-0_14

Patankar A, Patwardhan A (2016) Estimating the uninsured losses due to extreme weather events and implications for informal sector vulnerability: a case study of Mumbai, India. Nat Hazards 80(1):285–310. https://doi.org/10.1007/s11069-015-1968-3

Paudel Y (2012) A Comparative study of public-private catastrophe insurance systems: lessons from current practices. Geneva Pap Risk Insur Issues Pract 37(2):257–285. https://doi.org/10.1057/gpp.2012.16

Platzer EK, Knodt M (2023) Resilience beyond insurance: coordination in crisis governance. Environ Syst Decis. https://doi.org/10.1007/s10669-023-09938-7

Poussin C, Timoner P, Peduzzi P, Giuliani G (2025) Past and future trends in swiss snow cover: multi-decades analysis using the snow observation from space algorithm

PUPR (2017) Modul manajemen penanggulangan bencana pelatihan penanggulangan bencana banjir 2017. Pusat Pendidikan Dan Pelatihan Sumber Daya Air Dan Kontruksi, 77

Rentschler J, Salhab M, Jafino BA (2022) Flood exposure and poverty in 188 countries. Nat Commun. https://doi.org/10.1038/s41467-022-30727-4

Renn O, Laubichler M, Lucas K, Kröger W, Schanze J, Scholz RW, Schweizer PJ (2022) Systemic risks from different perspectives. Risk Anal 42(9):1902–1920

Rogers JC (1997) North Atlantic storm track variability and its association to the North Atlantic Oscillation and climate variability of northern Europe. J Clim 10(7):1635–1647

Roder G, Hudson P, Tarolli P (2019) Flood risk perceptions and the willingness to pay for flood insurance in the Veneto region of Italy. Int J Dis Risk Reduct 37:101172. https://doi.org/10.1016/J.IJDRR.2019.101172

Shah AA, Ajiang C, Khan NA, Alotaibi BA, Tariq MAUR (2022) Flood risk perception and its attributes among rural households under developing country conditions: the case of Pakistan. Water 14(6):992

Slovic P, Peters E (2006) Risk perception and affect. Curr Dir Psychol Sci 15(6):322–325. https://doi.org/10.1111/J.1467-8721.2006.00461.X

Surminski S, Eldridge J (2017) Flood insurance in England—an assessment of the current and newly proposed insurance scheme in the context of rising flood risk. J Flood Risk Manag 10(4):415–435. https://doi.org/10.1111/JFR3.12127

Tariq MAUR, Hoes OAC, Van de Giesen NC (2014) Development of a risk-based framework to integrate flood insurance. J Flood Risk Manag 7(4):291–307. https://doi.org/10.1111/JFR3.12056

Teriyoko W, Mudjiatko M, Sutikno S (2015) Identifikasi Wilayah Rawan Banjir Genangan Menggunakan Teknologi Sistem Imformasi Geografis (SIG) (Studi Kasus: Kota Pekanbaru) (Doctoral dissertation, Riau University)

Tesselaar M, Wouter Botzen WJ, Haer T, Hudson P, Tiggeloven T, Aerts JCJH (2020) Regional inequalities in flood insurance affordability and uptake under climate change. Sustainability (Switzerland) 12(20):1–30. https://doi.org/10.3390/su12208734

The World Bank (2021) How Indonesia strengthened its disaster response with risk finance and insurance. Feature Story The World Bank. https://www.worldbank.org/en/news/feature/2021/11/17/how-indonesia-strengthened-its-disaster-response-with-risk-finance-and-insurance

The World Bank (2022) World bank approves support to improve flood resilience and risk management in Indonesia. Press release no. 2023/038/EAP

Thistlethwaite J, Henstra D, Brown C, Scott D (2018) How flood experience and risk perception influences protective actions and behaviours among Canadian homeowners. Environ Manage 61(2):197–208

Varangis P, Skees J, Barnett B (2003) Weather indexes for developing countries. Forest 4:1

Yamamoto K, Sayama T, Apip (2021) Impact of climate change on flood inundation in a tropical river basin in Indonesia. Prog Earth Planet Sci 8(1):1–15. https://doi.org/10.1186/S40645-020-00386-4/FIGURES/12

Zinda JA, Williams LB, Kay DL, Alexander SM (2021) Flood risk perception and responses among urban residents in the northeastern United States. Int J Disaster Risk Reduct 64:102528

7 Indonesia's Foreign Policy on the Status of Forces Agreement (SOFA) in Post Natural Hazard-Induced Disasters Contexts

R. R. Emilia Yustiningrum and Lili Romli

Abstract

From the presidency of Sukarno (1945–1966) to Megawati (2001–2004), the management of natural hazard-induced disasters in Indonesia was predominantly treated as a domestic political issue. However, the devastating 2004 Aceh tsunami marked a turning point, prompting the Indonesian government to incorporate natural hazard-induced disasters into its foreign policy agenda. Beginning with the administration of President Susilo Bambang Yudhoyono (2004–2014) and continuing under President Joko Widodo (2014–2023), Indonesian foreign policy began to embrace the idea of foreign military involvement in post-disaster response. This chapter employs a qualitative approach, collecting data from primary sources through interviews with foreign policymakers from the Ministries of Foreign Affairs and Defense, and secondary sources such as academic journals, books, and reports from governmental and international organizations. Using the analytical framework of the Status of Forces Agreement (SOFA) in post-disaster contexts, the chapter argues that by incorporating natural hazard-induced disasters into its foreign policy agenda, the Indonesian government has opened the door to involving foreign military forces in disaster response within its territory.

Keywords

Indonesia · Foreign policy · Status of forces agreement (SOFA) · Natural disaster

R. R. E. Yustiningrum (✉) · L. Romli
Research Center for Politics BRIN, Jakarta, Indonesia
e-mail: emilia.yustiningrum@gmail.com

7.1 Introduction

Indonesia is a natural disaster-prone area located in the Ring of Fire that makes it vulnerable to various hazards in cyclones, droughts, earthquakes, extreme temperatures, floods, landslides, volcanic eruptions, and wildfires. Since the formation period, Indonesia suffered numerous disasters throughout its governments. During the Sukarno era, the First President administration (1945–1966), there were significant natural hazard-induced disasters including Tanimbar cyclone 1961 (Bullard 2017), the Java flood 1961 (Bullard 2017), the Flores earthquake 1961 (Northrop and Raitt 1963), and Mount Agung volcanic eruption 1963 (Zen and Hadikusumo 1964). Suharto's government faced large-scale disasters on Irian Jaya (now West Papua) earthquake in 1976 (Hamzah et al. 2000), the Bali earthquake in 1976 (Leimena 1979; McCaffrey and Nabelek 1987), the Flores earthquake and tsunami

D. Hidayati et al. (eds.), *Examining Disaster Risk Reduction in Indonesia*,
Sustainable Development Goals Series, https://doi.org/10.1007/978-981-96-4182-6_7

in 1992 (Kawata et al. 1995), Biak earthquake 1996 (Henry and Das 2002), and Irian Jaya drought 1998. The three short-lived governments of Baharudin Jusuf Habibie, Abdurrahman Wahid, and Megawati Sukarnoputri colored with fewer natural hazard-induced disasters in the country. However, the beginning of SBY's first presidential term marked the largest natural disaster in Indonesia's political history—the Aceh tsunami in 2004.

Indonesia's initial national regulations to manage disasters was managing the wounds and the sick during the independence war against the Dutch colonizer. It came partly about because Sukarno's government was heavily recovered from being colonized into an independent country. The arrival of Suharto in the presidential office in the 1960s transformed the national regulations to manage disaster risk into the distribution of humanitarian aid to the affected areas (Rozi and Yustiningrum 2016). Suharto's departure from the presidential office and the arrival of newly democracy in Indonesia in 1998 marked the emergence of various ethnic and religious conflicts. These led the three succeeding governments (Baharudin Jusuf Habibie, Abdurrahman Wahid, and Megawati Sukarnoputri) to transform the national regulations to manage natural hazards not only in the distribution of humanitarian assistance to the damaged regions but also in the administration of internally displaced people (IDPs) in the local conflict areas. Until the early first presidential term of SBY in 2004, with the exception of 1992 tsunamigenic earthquakes in Flores, Indonesia's national regulations to manage naturally hazard-induced disasters were still largely domestic issues that are barely governed by the involvement of foreign entities, either multinational militaries or international organizations.

The Aceh tsunami that happened on December 26, 2004 has changed the Indonesian government's policy on engaging foreign militaries in post-natural hazard-induced disasters context. The magnitude, the immense losses and damages, the decision of the SBY government to declare it as a state of emergency, the consequences afterward, the request to ask for help from the international community through targeted assistance, the arrangement to open Aceh that was still in the middle of intrastate conflict, the understanding that the only state agency who was able to manage the tsunami during the emergency period was the military either from home or abroad, and the agreement to administer the incoming of foreign militaries and humanitarian workers from various international organizations, were the significant foundations on changing of Indonesia's natural disaster management policies. For the first time in modern Indonesian political history, it opened the door for the incoming of foreign militaries in post-disaster periods in its territories.

Indonesia is still a member of the non-alignment movement and has never been a joiner of any security dialogues or military alliances worldwide. Its active and independent foreign policy principle has made the Indonesian government against any military enclaves in its territory. It happened partly because Indonesia used to be a colonized country and any external interference would have immense impacts on domestic politics. It should be acknowledged that the incoming of foreign militaries in Indonesia's national territories in the aftermath of natural hazard-induced disasters has to be scrutinized on the contemporary defense and foreign policies. Despite being immensely a non-alignment movement supporter, Indonesia slightly changed its foreign policy to create temporary agreements with foreign countries on stationing their militaries in its territories due to the enormous impacts of disasters in domestic politics.

This chapter tries to fill in the gap on how the Indonesian government changed its national regulations on managing natural hazard-induced disasters by allowing the incoming of multinational troops in the emergency period. By focusing on the periods of both President SBY (2004–2014) and Joko Widodo (2014–2023), the chapter tries to shed light on the temporary agreement on the stationing of foreign military in Indonesia's territories in post-natural hazard-induced disaster events known as Status of Forces Agreement

(SOFA) as an integrated part of its contemporary defense and foreign policies.

7.2 Analytical Framework

Status of Forces Agreement (SOFA) is an agreement that regulates the status of foreign military forces stationed in the territory of another country. These agreements usually cover aspects such as the rights and obligations of foreign troops, legal responsibilities, and troop withdrawal procedures. There are many definitions of SOFA. Fleck (2009) stated that SOFA is the legal framework that defines the rights and obligations of a foreign visiting force in a receiving state's territory.

Sari (2008) mentioned that SOFA is a bilateral or multilateral treaty that defines the legal position of military forces and civilian personnel deployed by one or more states. SOFAs typically deal with issues such as the entry and the departure of foreign personnel; the carrying of arms, taxation, the settlement claims; and the modalities for the exercise of civil and criminal jurisdiction over members of the visiting forces or mission (Sari 2008, pp. 67–68).

SOFA usually covers aspects such as the purpose and scope of the presence of foreign military forces, the rights and obligations of foreign military forces, legal responsibility for the actions of foreign military forces; and procedures for the withdrawal of foreign military forces. The types and contents of agreements in SOFA can vary, depending on the mission of the recipient countries. Therefore, the contents of each SOFA may differ from one country to another. In another word, there is no uniform international practice regarding the mutual agreement (SOFA) between the sending and receiving countries to deploy troops (Fleck 2009).

There are no formal requirements regarding the form, content, length, or title of a SOFA. A SOFA can be written for a specific purpose or activity, or it can anticipate a long-term relationship and provide maximum flexibility and applicability. It is generally a stand-alone document created as an executive agreement. SOFAs can include many provisions, but the most commonly discussed issue is which countries can exercise criminal jurisdiction. Other requirements found in SOFA include, but are not limited to, wearing uniforms, taxes and fees, carrying weapons, use of radio frequencies, licensing, and customs regulations (Mason 2009).

In general, Fleck (2009) divides SOFA into three categories: SOFAs for Peace Operations; SOFAs for Military Cooperation, and SOFAs for Unilateral Use of Training and Exercise. Table 7.1 shows the types of SOFA according to Fleck (2009).

Table 7.1 Types of SOFAs (*Source* Fleck 2009, p. 9)

SOFAs for peace operations	Time-limited but renewable mandates
SOFAs for military cooperation	Long-term (alliance) activities for general purposes or ad hoc missions for specific cooperation projects
SOFAs for a unilateral use of training and exercise	Facilities single or recurrent missions for unilateral use of specific facilities

Although the contents of SOFA vary from one country to another, one thing must be considered, namely, respect for the law of the receiving state and concurring jurisdiction. Fleck (2009) said it is the duty of members of the visiting force and its civilian members, as well as their dependents: to respect the law of the receiving state and—to abstain from any activity inconsistent with the force's mandate, especially political activity in the receiving state. Besides that, both the sending and the receiving state can exercise jurisdiction in what is called concurrent jurisdiction. The three following provisions should be included in a SOFA: (1) a description of situations for which the receiving state may exercise jurisdiction; (2) a description of other situations for which the sending state may exercise jurisdiction; and (3) a need for cooperation between competent authorities of the sending and receiving state (Fleck 2009, pp. 16–18).

Many actors are, either ministries or institutions, involved in the policymaking of SOFA.

The main actors are the Ministry of Defence and the Ministry of Foreign Affairs. However, these two ministries are insufficient, but must also involve other ministries or institutions related to the problem being handled, such as the Ministry of Home Affairs and the Police. When it relates to natural hazards, it is necessary to involve institutions that deal with natural hazard-induced disasters. Several phases must be passed before the SOFA is approved. Fleck (2009, p. 24) mentioned that five phases must be passed in negotiations to reach an agreement, namely: initial contact between the sending and the receiving states; initial contact between the sending and the receiving states; elaboration of a first draft of the SOFA; formal negotiations; and approval process, that can be seen in Table 7.2.

Initially, SOFA was a military assistance agreement made by the Allied Countries during World War I. During that period, SOFA was further developed in and after World War II, leading to, among other things, NATO SOFA and crisis management SOFA. Then, after World War II, international tensions led to the start of the Cold War, which resulted in the semi-permanent deployment of large numbers of extraterritorial troops in both the East and West. The North Atlantic Treaty Organization (NATO) was founded in Western Europe in 1949. To regulate the presence of the armed forces of member states on each other's territory within the framework of this alliance, a new treaty was negotiated based on the SOFA draft. In later developments, the UN also sent peace troops based on SOFA to countries that needed the assistance of UN peace troops. Likewise, other international organizations are involved in crisis management operations, such as the African Union and the European Union (Voetelink 2022).

In the context of natural hazard-induced disasters, SOFA can be used to organize the presence of foreign military forces to provide humanitarian assistance. This agreement can provide legal certainty and security for foreign troops involved in humanitarian assistance efforts and protect the rights and interests of host country citizens. SOFA is needed for this military assistance because there is a tendency for armed forces around the world to go beyond traditional warfare and carry out tasks related to humanity and development. Some factors behind this are post-Cold War realignment, professionalization of the armed forces (phasing out conscription and greater investment in the training and pay of individual soldiers) and the search for a new role as a 'force for good' or 'warrior of humanity.' (Wheeler and Harmer 2006; Wiharta et al. 2008). It also reflects a move toward a more comprehensive approach to security. One of the key variables influencing the policies of countries contributing to the delivery of military assets to SOFA is national strategic culture, which relates to the perception and actual role of the military in society and the world. Beliefs about the proper role of armed forces greatly influence the missions they are required to undertake. In some countries, it is considered normal for the army to play a central role in responding to natural or human-caused disasters (Wiharta et al. 2008).

Militaries deployed to assist with natural hazard-induced disasters should refer to the Oslo Guidelines, formulated in 1994. These guidelines are intended to address the need for principles and standards and to provide improved coordination in using military and civil defense assets in response to natural, technological, and environmental emergencies in peacetime (Homan 2008, October 1–2). The transition is that aid is based on actual needs, provided by actors with no political interests, and contributes to the safety of survivors of natural hazard-induced disasters. The Oslo Guidelines were subsequently updated in 2006. These revisions were partly influenced by the 2003 Guidelines on the Use of Military and Civil Defence Assets to Support the United Nations (Wiharta et al. 2008, p. 10).

According to Wiharta et al. (2008, pp. xii–xiii), there are six interconnected aspects of the effectiveness of the use of foreign military assets, namely: timeliness, appropriateness, efficiency, absorption capacity, coordination, and cost. Of these six aspects, the chapter quotes directly from the executive summary, namely:

> *Timeliness* seems to be the main factor affecting the effectiveness of foreign military assets in a natural disaster response, especially in the first

Table 7.2 Organization of SOFA negotiations (*Source* Fleck 2009, p. 24)

Phase 1: Initial contact between the sending and the receiving states	Phase 2: Clarification of needs regarding the presence of the visiting force	Phase 3: Elaboration of a first draft of the SOFA	Phase 4: Formal negotiations	Phase 5: Approval process
Actors: Foreign and defence ministries	Actors: Foreign and defence ministries	Actors: Foreign and defence ministries, coordination with other ministries	Actors: Foreign and defence ministries as well as further ministries, such as interior, finance, justice, environment	Actors: All concerned ministries, parliament
Important issues: Which other actors have to be informed? E.g., Parliament, other ministries, media, etc	Important issues: Which types of concerns are to be anticipated? E.g., Political, legal or constitutional, economic, etc	Important issues: Who is in charge? Who has to be consulted?	Important issues: Are competing interests involved? How to reconcile competing interests?	Important issues: Is parliamentary approval required? Does it involve legislative acts? Is a provisional application feasible?

days and weeks of the operation. In particular, military aircraft can transport large quantities of relief supplies and other assets and military helicopters can support search-and-rescue operations. However, when promised military assets are slow to arrive and to start operating it may actually impede the response by preventing the deployment of civilian alternatives. The timely arrival of foreign military assets can be affected by their location at the time of the disaster and bureaucratic delays relating to, for example, status-of-forces agreements (Wiharta et al. 2008).

The appropriateness of a military asset is determined by how well its capabilities meet the needs of the response and how suitable it is for the local cultural and political context in which it operates. The study highlights the importance of needs assessments. Regular comprehensive, multi-stakeholder needs assessments, linked to a coordination framework, can help to ensure that the appropriate assets—military and civilian—are provided when they are needed and to facilitate the withdrawal of assets that are no longer required or appropriate (Wiharta et al. 2008).

The efficiency of a foreign military asset in a natural disaster response depends not only on the efficiency with which it carries out its assigned tasks but also on how well its capabilities are used within the more extensive operation. The former aspect is, to a large extent, affected by the techniques used; the latter is related to the coordination of the relief operation and to how far foreign military contingents submit to coordination by other, often civilian, actors. Some countries that contribute military assets often insist on force-protection measures, which both reduce the efficiency of the operation and may intimidate or be resented by local populations (Wiharta et al. 2008).

The effectiveness of foreign military assets in disaster relief is also affected by absorptive capacity in the affected country—the ability of disaster management institutions to coordinate and effectively use the assets during the relief operation. While individual military assets tend to be relatively self-sufficient and thus to place a small burden on absorptive capacity, the arrival of large numbers of foreign military assets from different countries and with overlapping capabilities can cause serious problems. (Wiharta et al. 2008, p. xii).

Coordination between civilian humanitarian actors and military assets has been one of the greatest challenges created by the increasing deployment of foreign military assets (Homan 2008, October 1–2). The differences in cultures, priorities and operating modes between military personnel and civilian actors have an impact not least on information sharing between the civilian and military spheres (Parisetti 2012). Information management is crucial to the success or failure of any relief operations. This role is best and most suitably carried out by the United Nations, led by OCHA (Wiharta et al. 2008).

> *The costs* of deploying military assets are generally higher than for civilian assets (Homan 2008, October 1–2; Parisetti 2012). This has caused concerns that foreign military assets are placing a disproportionate burden on humanitarian funds. Several countries have introduced measures whereby their defence ministry covers some or all of the costs of deploying military assets for overseas disaster relief, reducing their impact on humanitarian aid budgets. The implications of this for humanitarian funds at the international level are hard to gauge in the absence of greater transparency in reporting (Wiharta et al. 2008).

As military aid increases, civilian humanitarian actors often worry about being too closely aligned with military forces. Several issues emerged, such as: interactions between civilian and military actors in disaster situations; how to maintain humanitarian space based on neutrality, impartiality, and humanity; and the conditions under which military assets should be deployed and when they are inappropriate. The following are several things that need to be considered in SOFA for natural hazard-induced disasters (Fleck 2009; Wiharta et al. 2008). First, aims and scope of humanitarian assistance. Agreements should clearly define the purpose and scope of humanitarian assistance to be provided by foreign forces. It is important to ensure that foreign forces only engage in activities that comply with international law and the host country's national laws (Fleck 2009; Wiharta et al. 2008).

Second, rights and obligations of foreign troops. The treaty should set out the rights and obligations of foreign troops involved in humanitarian assistance. This is important to protect the rights and interests of foreign troops and prevent violations of the law (Fleck 2009; Wiharta et al. 2008). Third, legal responsibility. The treaty should provide for legal responsibility for the actions of foreign forces involved in humanitarian assistance. This is important to ensure that survivors of disasters receive adequate compensation if legal violations occur (Fleck 2009; Wiharta et al. 2008). Fourth, troop withdrawal procedures. The agreement should set out procedures for the withdrawal of foreign troops involved in humanitarian assistance. This is important to ensure that the withdrawal of troops is carried out orderly and smoothly (Fleck 2009; Wiharta et al. 2008).

7.3 Discussion

7.3.1 Natural Hazard-Induced Disasters in Indonesia (2004–2022)

Indonesia is an area that is vulnerable and at high risk of natural hazard-induced disasters due to geographical, geological, and hydrological conditions and is located at the meeting point of three world tectonic plates: Eurasia, Indo-Australia, and the Pacific. The tectonic activity that occurred caused the formation of a series of volcanoes along the islands of Sumatra, Java-Bali-Nusa Tenggara, north of Sulawesi-Maluku, to Papua, which is part of a series of volcanoes throughout the Asia–Pacific as the Ring of Fire so that earthquakes often occur. Tectonic activity also forms faults, such as the Semangko fault in Sumatra, the Sorong fault in Papua and Maluku, and the Palukoro fault in Sulawesi (Amri et al. 2016). The distribution of faults is closely related to the distribution of earthquake centers. The earthquake and tsunami that occurred were heavily influenced by tectonic fault activity (Amri et al. 2016).

Natural hazard-induced disasters in Indonesia are not only caused by tectonic activity or geological factors, but also by hydrometeorological factors, namely in the form of floods, extreme waves, land and forest fires, drought, and extreme weather. Floods can be caused by static natural conditions such as geography, topography, and river channel geometry. Flash floods have the potential to occur in river basin areas that are formed from hilly valleys with steep slopes and have abundant water sources. This river basin area will also become more vulnerable if there is a lot of material supporting landslides and river blockages. Land and forest fires are not natural hazard-induced disasters, because 99% of incidents in Indonesia are caused by human factors, either intentionally or through negligence (Amri et al. 2016). Extreme weather is related to extraordinary events that have the potential to cause disaster, including tropical cyclones, and tornadoes (Amri et al. 2016; Kogan 1997). Droughts are becoming increasingly severe and are reported every 3 years during the dry season. This drought disaster mostly occurred on the southern islands of Java, Bali, and Nusa Tenggara. This disaster will negatively impact food security, clean water availability, and water deficits, as happened in Bali and East Nusa Tenggara (Perwaiz et al. 2020).

In this regard, the Asian Disaster Preparedness Center (ADPC) stated that due to Indonesia's location on the Pacific Ring of Fire, many active volcanoes are found in the country (CFE-DM 2018; Perwaiz et al. 2020). For example, Mount Agung has continued activity from late 2017 until June 2019, generating frequent eruptions and generous amounts of tephra. Prevalence of such hazards continues to endanger the population, infrastructure, and development. Due to the prevalence and high frequency of the hazards Indonesia is exposed to and vulnerable to, the impacts of disasters may significantly hinder the economy and its future development. Earthquakes have also caused significant damage to the country's economic performance. Between 1970 and 2015, there were 97 large earthquakes, which amounted to US$ 11.7 billion in losses and damages (CFE-DM 2018; Perwaiz et al. 2020, pp. 6–8).

Based on data from the Indonesian National Agency for Disaster Management (BNPB), there were 42,827 natural hazard-induced disaster events in Indonesia from 2005 to 2023, divided into 41,976 hydrometeorological disasters and 851 geological disasters (BNPB, 2024). For details of the two disasters, see the Table 7.3.

Table 7.3 Number of hyrometeorologinal and geological disasters in Indonesia, 2005–2023 (*Source* BNPB, 2024)

No.	Disasters	Numbers
1	Flood	13,973
2	Tornado	11,837
3	Landslide	9844
4	Forest Fire	4461
5	Drought	1861
6	Earthquake	615
7	Volcanic Eruption	199
8	Tsunami	37
Total		42,827

In response to the significant number of natural hazard-induced disasters, BNPB conducted a study on risk focusing on earthquake- and tsunami-related hazards. Table 7.4 presents data on the extent of exposure to earthquake disaster risk in the province, summarizing medium-to-high disaster risk levels. Similarly, Table 7.5 provides a matrix detailing the extent of exposure to tsunami disaster risk in the province, also summarizing medium-to-high disaster risk levels.

From 2004 to 2022, geological hazards such as earthquakes and tsunamis caused significant casualties and widespread impact. One of the most devastating events occurred on December 26, 2004, in Aceh. A 9.0 magnitude earthquake struck beneath the Indian Ocean seabed, approximately 200 km off the coast of Banda Aceh, Sumatra (Wiharta et al. 2008). The earthquake triggered a massive tsunami that affected 12 countries around the Indian Ocean (Wiharta et al. 2008). Traveling at speeds of 600–800 km per hour in open water, the tsunami waves reached heights of up to 20 m as they struck the coast at speeds of up to 60 km per hour (Wiharta et al. 2008). In Aceh province, the tsunami devastated a 500-km stretch of coastline and swept nearly 5 km inland (Wiharta et al. 2008).

The disaster claimed approximately 125,866 lives, displaced around 419,682 people, and caused damages and losses amounting to 4.45 billion USD (Wiharta et al. 2008). Infrastructure was severely impacted, with the road connecting Meulaboh to Medan and the provincial capital Banda Aceh collapsing. Many local officials, along with their family members, either perished or went missing, leaving regional governments, including provincial authorities, struggling to function effectively. In response, Indonesian President Susilo Bambang Yudhoyono declared a 3-month state of emergency (Wiharta et al. 2008).

The Yogyakarta earthquake occurred on May 27, 2006, with a magnitude of 5.9 on the Richter scale and lasted for 52 s. Due to its relatively shallow depth, the earthquake caused significant destruction, particularly in Bantul District, Yogyakarta Province, and Klaten, Central Java Province. The disaster resulted in an estimated

Table 7.4 Number of exposures to earthquake disaster risk in the province (medium–high disaster risk recapitulation) (*Source* Amri et al. 2016, p. 59)

Province	Social (people)	Physical (IDR million)	Economy (IDR. million)
Aceh	3,695,3999	35,182,536	12,170,729
North Sumatra	9,882,415	49,509,931	36,531,257
West Sumatra	4,940,818	37,965,427	22,311,824
Riau	719,788	2,652,365	5,443,724
Jambi	560,085	3,286,943	3,443,724
South Sumatra	1,003,680	4,416,456	831,373
Bengkulu	1,829,382	12,469,145	5,695,270
Lampung	1,577,979	6,023,539	3,745,855
DKI Jakarta	852,831	2,320,653	31
West Java	15,347,975	53,502,015	9,445,160
Central Java	8,736,975	34,714,529	10,741,733
Yogyakarta	2,575,521	11,101,607	2,474,748
East Java	8,617,961	27,320,393	9,993,782
Banten	5,479,036	15,982,134	4,273,501
Bali	1,532,905	8,464,733	388,401
West Nusa Tenggara	3,666,884	17,930,957	4,915,045
East Nusa Tenggara	1,375,285	8,728,706	2,076,253
North Kalimantan	5,067	27,554	63,289
North Sulawesi	2,023,445	12,596,855	5,029,826
Central Sulawesi	2,256,213	16,727,968	13,579,841
South Sulawesi	1,919,957	11,399,854	11,946,899
Southeast Sulawesi	506,675	3,021,500	2,745,928
West Sulawesi	1,058,857	6,655,380	3,553,796
Gorontalo	1,123,053	8,164,378	3,553,796
Maluku	953,583	12,764,393	1,949,189
North Maluku	1,007,227	10,594,807	920,025
West Papua	576,481	11,079,332	710,213
Papua	2,420,778	42,085,684	1,540,624
Total	86,247,258	466,689,834	182,185,171

death toll of 5716 people (BAPPENAS 2006). Damage to housing was extensive, exceeding 50% of the total structures in the affected areas (BAPPENAS 2006). Private homes suffered the most, accounting for more than half of the total damage and losses, which amounted to IDR 15.3 trillion (BAPPENAS 2006). Additionally, private sector buildings and productive assets were heavily impacted, with losses estimated at IDR 9 trillion (BAPPENAS 2006).

On September 30, 2009, a 7.6-magnitude earthquake struck West Sumatra, with its epicenter located approximately 50 km northwest of Padang City. The earthquake impacted multiple areas, including Padang City, Padang Pariaman Regency, Pariaman City, Pesisir Selatan Regency, Bukit Tinggi City, Padangpanjang City, Solok City, Agam Regency, and West Pasaman Regency (Yustiningrum 2016a). The disaster resulted in 1117 fatalities, 1214 serious injuries, and 1688 minor injuries (BPBDKP 2019, February 4). Additionally, 135,448 houses were heavily damaged, 65,380 moderately damaged, and 78,604 slightly damaged (BPBDKP 2019,

Table 7.5 Number of tsunami disaster risk exposures in the province (medium–high disaster risk recapitulation) (*Source* Amri et al. 2016, p. 65)

Province	Social (people)	Physical (IDR million)	Economy (IDR million)	Environment (ha)
Aceh	294,623	7,499,417	564,258	5618
North Sumatra	124,544	1,557,294	460,853	11,303
West Sumatra	272,020	4,042,000	340,368	14,765
Bengkulu	77,888	530,356	511,212	181
Lampung	29,369	85,355	73,202	191
DKI Jakarta	15,259	839,383	591,824	59
West Java	127,007	729,531	291,948	148
Central Java	91,864	2,677,842	359,859	1005
Yogyakarta	36,893	90,112	1,484,518	–
East Java	176,223	3,205,659	915,737	1896
Banten	205,745	981,501	254,194	1608
Bali	179,692	3,182,547	252,308	350
West Nusa Tenggara	348,629	9,252,560	630,347	6732
East Nusa Tenggara	114,272	2,892,049	349,648	8618
North Sulawesi	199,092	3,450,744	132,258	2817
Central Sulawesi	143,067	5,227,777	118,214	6837
South Sulawesi	399,555	4,924,338	344,052	2591
Southeast Sulawesi	40,208	136,696	720	1446
Gorontalo	8499	411,063	2819	624
West Sulawesi	85,550	675,354	70,836	3340
Maluku	389,917	5,800,421	25,532	18,308
North Maluku	215,182	5,266,984	199,152	11,679
West Papua	62,573	2,680,884	1236	8877
Papua	65,031	5,354,954	1265	9695
Total	3,702,702	71,494,821	7,976,358	119,688

February 4). The total damage and losses amounted to IDR 21.5 trillion (Yustiningrum 2016a).

On October 25, 2010, another disaster struck the Mentawai Islands off the coast of West Sumatra when a 7.2-magnitude earthquake triggered a tsunami. The earthquake's epicenter was located on the seabed, around 78 km south of Pagai Island and approximately 10 km below sea level (Purnama 2023, October 25). The severe shaking was felt across nearby islands, including Siberut and Pagai. According to BNPB data, the event caused 509 fatalities, injured 24 people (both seriously and lightly), left 21 missing, and displaced 11,425 individuals (Purnama 2023, October 25). The disaster also resulted in damage and losses totaling IDR 348.92 billion (Purnama 2023, October 25).

The eruption of Mount Merapi on October 27, 2010. This eruption of Merapi is not the first, previously, there were large eruptions in 1822, 1872, and 1930–1931. The Merapi eruption that occurred in 2010 resulted in 356,816 people being displaced and 270 people dying (Yustiningrum 2016b). The affected population came from Sleman Regency, Magelang Regency, Klaten Regency, and Boyolali Regency. Other data states that around 240 residents living around Mount Merapi died as a result of being hit by hot clouds (Yustiningrum 2016b). The damage and losses resulting from the eruption of Mount Merapi amounted to IDR 3.62 trillion (Andryanto 2021b, October 26). The details of the damage are, the productive economic sector amounted to IDR 1.69 trillion, infrastructure IDR 707.42 billion, housing IDR 626.65 billion, cross-sectors IDR 408.75 billion, and the social sector IDR 122.47 billion (Andryanto 2021a, b, October 26).

Lombok and Sumbawa earthquakes occurred four times in 2018. The first earthquake occurred on July 29, 2018, with a magnitude of 6.4 on the Richter scale. On August 5, 2018, another earthquake with a magnitude of 6.9 on the Richter scale occurred. Then on August 9, 2018, another earthquake occurred with a magnitude of 5.9 on the Richter scale. And on August 19, 2018, the fourth earthquake occurred again with a magnitude of 6.0 on the Richter scale. The earthquake shock was felt in West Lombok, East Lombok, North Lombok, Central Lombok, West Sumbawa, and Sumbawa Besar. From BNPB data, the 2018 Lombok earthquake damaged 71,962 housing units, 671 educational facilities, 52 health facilities, 128 worship facilities, and infrastructure facilities (Rachmawati 2021, August 6). Meanwhile, casualty data shows that 460 people died, 7733 people were injured, 417,529 people were displaced (Rachmawati 2021, August 6). The total loss and economic damage was more than US$ 530 million (Rachmawati 2021, August 6).

On September 28, 2018, a devastating earthquake and tsunami struck Central Sulawesi. The earthquake, with a magnitude of 7.7 on the Richter scale, was centered at 0.18° South Latitude and 119.85° East Longitude, approximately 27 km northeast of Donggala (Septiyan 2023, September 9). The disaster severely impacted several areas, including Palu City, Sigi Regency, Donggala Regency, and Parigi Moutong Regency. According to UN records, the disaster claimed at least 4845 lives, displaced 172,999 people, and damaged 110,214 houses (Septiyan 2023, September 9). The total losses from this multidimensional disaster reached IDR 18.48 trillion, with Palu City suffering losses of IDR 8.3 trillion, Sigi Regency IDR 6.9 trillion, Donggala Regency IDR 2.7 trillion, and Parigi Moutong Regency IDR 640 billion (Septiyan 2023, September 29).

On December 22, 2018, another disaster struck when the Tanjung Lesung tsunami occurred. Triggered by the eruption of Anak Krakatau in the Sunda Strait, this event affected coastal areas in Banten and Lampung. According to government reports, the tsunami occurred four times in four different locations, with tidal waves reaching heights of 0.3–0.9 m (Andryanto 2021b, December 22). At least 426 people were killed, 7202 were injured, and 23 were reported missing (Andryanto 2021a, b, December 22). The National Disaster Management Authority (BNPB) noted that the tsunami caused damage to 2752 houses, 92 accommodation and food stalls, 510 boats and ships, and 147 vehicles (Farisa 2018, December 31).

On November 21, 2022, a magnitude 5.6 earthquake struck Cianjur, West Java. The quake was caused by movement along the Cugenang Fault, which stretches approximately 9 km and crosses at least nine villages: Ciherang, Ciputri, Cibeureum, Nyalindung, Mangunkerta, Sarampad, Cibulakan, and Benjot (Siswadi 2023, November 21). These villages are located in the Cugenang and Cianjur sub-districts. The earthquake resulted in 635 fatalities, 2046 injuries, and damage to thousands of buildings (Siswadi 2023, November 21). Total losses due to this earthquake were estimated at around IDR 4 trillion (Siswadi 2023, November 21).

7.3.2 The Involvement of Foreign Military in Natural Hazard-Induced Disasters

Wiharta et al. (2008: 92–94) detailed the role of foreign militaries in assisting during the 2004 Aceh tsunami disaster. The militaries involved included the Australian Defence Force (ADF), the Singapore Armed Forces (SAF), the US Pacific Command (USPACOM), and Japan, which deployed units from its Air Self-Defence Force (ASDF), Maritime Self-Defence Force (MSDF), and Ground Self-Defence Force (GSDF). Additionally, military contingents from 10 other countries supported Indonesia's TNI during the 3-month emergency period.

The Australian Defence Force provided significant assistance, including deploying a C-130 Hercules aircraft carrying medical supplies and aid, as well as setting up a 34-bed field hospital (Wiharta et al. 2008). The HMAS Kanimbla served as a medical facility and a sea base for equipment and personnel. The ADF delivered approximately 1200 tons of emergency humanitarian aid, including food, water, medical supplies, and shelter equipment, with a primary focus on health, water, and sanitation (Wiharta et al. 2008). Most of the aid was directed to Banda Aceh, with smaller-scale support reaching the northwest coast and remote islands such as Nias, Batu, Banyak, and Simeulue.

The Singapore Armed Forces contributed with C-130 Hercules aircraft, Chinook and Super Puma helicopters, 130 medical personnel, and 103 engineers (Wiharta et al. 2008). SAF helicopters maintained an air bridge between Medan, Meulaboh, and Banda Aceh, facilitating the delivery of relief supplies and the evacuation of injured survivors. Similar to HMAS Kanimbla, the SAF's Landing Ship Tank (LST) fleet allowed personnel and equipment to operate at sea, bolstering logistics and transport capabilities for the relief effort.

Military assistance from the US via the aircraft carrier USS Abraham Lincoln. Three other ships—USS Bonhomme Richard, USS Essex, and USS Fort McHenry—supported relief operations for a month starting in early January 2005. Twenty-eight helicopters from the ships flew more than 100 missions each day, dropping relief supplies and evacuating people who needed medical assistance from communities on the west coast of Aceh (Wiharta et al. 2008). The hospital ship USNS Mercy also joined in the rescue operation for tsunami survivors.

The Japanese Minister of Defense ordered the deployment of ASDF airlift units, maritime units of the MSDF, and medical units of the GSDF, as well as joint liaison officers of the Joint Staff Office. Two C-130 Hercules arrived in Banda Aceh from Utapao carrying relief supplies. A GSDF advance team consisting of 20 personnel provided medical treatment 3 days later (Wiharta et al. 2008). GSDF provides medical and sanitation support, vaccination, and other public health measures to prevent disease outbreaks. Three ships supporting the GSDF operation transported relief materials to the southwest of the provincial capital. Its hovercraft transported cleaning vehicles to remote areas along the coast, supporting TNI efforts to clear roads ahead of repair work that will re-establish road links between Banda Aceh and southwestern districts.

References

Amri MR, Yulianti G, Yunus R, Wiguna S, Adi AW, Ichwana AN, Septian RT (2016) RBI. Risiko Bencana Indonesia. Badan Nasional Penanggulangan Bencana (BNPB), Jakarta, Indonesia

Andryanto SD (2021a, Dec 22) 22 Desember 2018, Bencana Tsunami Tanjung Lesung 426 Orang Tewas. Tempo.co. Retrieved from https://tekno.tempo.co/read/1542244/22-desember-2018-bencana-tsunami-tanjung-lesung-426-orang-tewas

Andryanto SD (2021b, Oct 26) Erupsi Dahsyat Gunung Merapi 11 Tahun Lalu, Mbah Maridjan Salah Seorang Korban. Tempo.co. Retrieved from https://nasional.tempo.co/read/1521393/erupsi-dahsyat-gunung-merapi-11-tahun-lalu-mbah-maridjan-salah-seorang-korban

BAPPENAS (2006) Penilaian Awal Kerusakan Dan Kerugian Bencana Alam Di Yogyakarta dan Jawa Tengah. Pertemuan ke 15 Grup Konsultatif untuk Indonesia. Badan Perencanaan Pembangunan Nasional (Indonesian Government), Jakarta, Indonesia

BNPB (2024) Tentang Data Informasi Bencana Indonesia (DIBI). National Agency on Disaster Management (Indonesian Government). https://dibi.bnpb.go.id/superset/dashboard/1/?standalone=0&expand_filters=0

BPBDKP. (2019, Feb 4) Mengenang Gempa 2009 Kota Padang. Local Agency for Disaster Management at the Padang City (BPBD Kota Padang or Local Government of Padang City). https://bpbd.padang.go.id/konten/mengenang-gempa-2009-di-kota-padang#:~:text=Menurut%20data%20Satkorlak%20PB%2C%20sebanyak,%2C%20%26%2078.604%20rumah%20rusak%20ringan

Bullard S (2017) In their time of need: Australia's overseas emergency relief operations, 1918–2006. Cambridge University Press, Cambridge

CFE-DM (2018) Indonesia disaster management reference handbook. Center for Excellence in Disaster Management and Humanitarian Assistance (CFE-DM)

Farisa FC (2018, Dec 31) Korban Meninggal Tsunami Selat Sunda Capai 437 Orang. Kompas.com. Retrieved from https://nasional.kompas.com/read/2018/12/31/15561231/korban-meninggal-tsunami-selat-sunda-capai-437-orang

Fleck D (2009) Guidebook drafting of status-of-forces agreements (SOFAs). Geneva Center for the Democratic Control of Armed Forces (DCAF), Geneva, Switzerland

Hamzah L, Puspito NT, Imamura F (2000) Tsunami catalog and zones in Indonesia. J Nat Dis Sci 22(1):25–43

Henry C, Das S (2002) The Mw 8.2, 17 February 1996 Biak, Indonesia, Earthquake: rupture history, aftershocks, and fault plane properties. J Geophys Res Solid Earth 107(B11):ESE 11-11-ESE 11-16. https://doi.org/10.1029/2001JB000796

Homan, K. (2008, Oct 1–2) Military support to civilian authority NATO's 2008 civil protection committee seminar: civil-military cooperation and interaction, Sibenik, Croatia

Kawata Y, Tsuji Y, Syamsudin AR, Sunarjo, Matsuyama M, Matsutomi H, Takahashi T (1995) Response of residents at the moment of Tsunamis—the 1992 flores Island Earthquake Tsunami, Indonesia. In: Tsuchiya Y, Shuto N (eds) Tsunami: progress in prediction, disaster prevention and warning. Springer, Dordrecht, pp 173–185

Kogan FN (1997) Global drought watch from space. Bull Am Meteor Soc 78(4):621–636. https://doi.org/10.1175/1520-0477(1997)078%3c0621:GDWFS%3e2.0.CO;2

Leimena SL (1979) Disaster in Bali caused by earthquake 1976 (a report). Disasters 3(1):85–87. https://doi.org/10.1111/j.1467-7717.1979.tb00203.x

Mason RC (2009) Status of forces agreement (SOFA): What is it, and how has it been utilized? Congressional research service 7-5700 RL 34551

McCaffrey R, Nabelek J (1987) Earthquakes, gravity, and the origin of the Bali Basin: an example of a Nascent Continental Fold-and-Thrust Belt. J Geophys Res Solid Earth 92(B1):441–460. https://doi.org/10.1029/JB092iB01p00441

Northrop J, Raitt RW (1963) Seaquakes in the Flores Sea. Bull Seismol Soc Am 53(3):577–583. https://doi.org/10.1785/bssa0530030577

Parisetti PC (2012) The Use of Civil and Military Defense Assets in Emergency Situations. In: de Guttry A, Gestri M, Venturini G (eds) International Disaster Response Law. T.M.C. Asser Press, pp 585–599. https://doi.org/10.1007/978-90-6704-882-8_24

Perwaiz A, Parviainen J, Somboon P, Mcdonald A, Kumar A, Wilcox T, Amach O (2020) Disaster risk reduction in the republic of Indonesia: status report 2020. United Nations Office for Disaster Risk Reduction (UNDRR), Regional Office for Asia and the Pacific and Asian Disaster Preparedness Centre (ADPC), Bangkok, Thailand

Purnama KI (2023, Oct 25) Kilas Balik Gempa Mentawai 25 Oktober 2010, Disertai Tsunami dan Ratusan Orang Meninggal. Tempo.co. Retrieved from https://nasional.tempo.co/read/1788679/kilas-balik-gempa-mentawai-25-oktober-2010-disertai-tsunami-dan-ratusan-orang-meninggal

Rachmawati R (2021, Aug 6) Mengenang Gempa Lombok 2018, Ratusan Orang Meninggal dan Ribuan Bangunan Rusak Berat. Kompas.com. Retrieved from https://regional.kompas.com/read/2021/08/06/083800578/mengenang-gempa-lombok-2018-ratusan-orang-meninggal-dan-ribuan-bangunan?page=all

Rozi S, Yustiningrum RE (2016) Memahami Erupsi Merapi: Kebijakan dan Implementasi. Graha Ilmu, Yogyakarta, Indonesia

Sari A (2008) Status of forces and status of mission agreements under the ESDP: the EU's evolving practice. Eur J Int Law 19(1):67–100. https://doi.org/10.1093/ejil/chn003

Septiyan MR (2023, Sept 29) Tsunami dan Gempa Palu Donggala 2018 dalam Angka: Korban, Daya Rusak, dan Lainnya. Tempo.co. Retrieved from https://nasional.tempo.co/read/1777905/tsunami-dan-gempa-palu-donggala-2018-dalam-angka-korban-daya-rusak-dan-lainnya

Siswadi S (2023, Nov 21) Setahun Gempa Cianjur 21 November, Gempa Susulan Terjadi 579 Kali. Tempo.co. Retrieved from https://tekno.tempo.co/read/1799654/setahun-gempa-cianjur-21-november-gempa-susulan-terjadi-579-kali

Voetelink J (2022) The status of forces agreements. In: Sayapin S, Atadjanov R, Kadam U, Kemp G, Zambrana-Tévar N, Quénivet N (eds) International conflict and security law: a research handbook. T.M.C. Asser Press, The Hague, pp 229–253

Wheeler V, Harmer A (2006) Resetting the Rules of Engagement: Trends and Issues in Military-Humanitarian Relations. Humanit Policy Group (HPG) Res Rep (21):1–88

Wiharta S, Ahmad H, Haine JY, Löfgren J, Randall T (2008) The effectiveness of foreign military assets in natural disaster response. Stockholm International Peace Research Institute (SIPRI), Sweden

Yustiningrum RE (2016a) Bencana Alam, Kerentanan, dan Kebijakan di Indonesia: Studi Kasus Gempa Padang dan Tsunami Mentawai. Graha Ilmu, Yogyakarta, Indonesia

Yustiningrum RE (2016b) Menuju Kebijakan Penanggulangan Bencana Letusan Gunung Merapi yang Berpihak pada Pengurangan Kerentanan terhadap Kemiskinan. In: Rozi S, Yustiningrum RE (eds) Memahami Erupsi Merapi: Kebijakan dan Implementasi. Yogyakarta, Graha Ilmu

Zen MT, Hadikusumo D (1964) Preliminary report on the 1963 Eruption of Mt. Agung in Bali (Indonesia). Bull Volcanol 27:269–299

8 The Role of Community Resilience in Disaster Risk Reduction

S. Andy Cahyono and Ngadisih

Abstract

Community resilience, a multifaceted concept, is analyzed through the lens of social science to provide a comprehensive understanding of its significance in mitigating the impacts of disasters. Key concepts such as social cohesion, adaptive capacity, and community-based strategies are examined in depth. The chapter highlights community resilience's dynamic and context-specific nature. The chapter also discusses the integration of community resilience into DRR policies and practices, emphasizing the importance of fostering local ownership and participation. Case studies from various disaster-prone regions are presented to illustrate how communities have harnessed their resilience to reduce vulnerability and enhance their capacity to respond to disasters effectively. This chapter underscores the need for a holistic approach to disaster risk reduction that recognizes the vital role of community resilience. It provides valuable insights for policymakers, practitioners, and researchers working in the field of DRR, offering practical recommendations for enhancing the resilience of communities and building a safer, more sustainable future in the face of mounting disaster threats.

Keywords

Community resilience · Disaster risk reduction · Adaptive capacity · Social cohesion · Community-based strategies

8.1 Introduction

Indonesia is a disaster-prone country that lies at the intersection between three tectonic plates, i.e., the Eurasian, Indo-Australian, and Pacific Ocean plates. This condition has made Indonesia known as a disaster supermarket (Nuryanti et al. 2017). According to BNPB (*Badan Nasional Penanggulangan Bencana*/National Disaster Management Agency), the most common disasters are volcanic eruptions, earthquakes, tsunamis, floods, and landslides. Not a few survivors fell in a kind of loss of life and property. The number of survivors depends on the perspectives of disaster management, whether it is more concentrated during and

S. A. Cahyono (✉)
Research Center for Ecology and Ethnobiology, National Research and Innovation Agency (BRIN), Cibinong, Indonesia
e-mail: sand009@brin.go.id

Ngadisih
Department of Agricultural and Biosystem Engineering, Faculty of Agricultural Technology, Gadjah Mada University (UGM), Yogyakarta, Indonesia

D. Hidayati et al. (eds.), *Examining Disaster Risk Reduction in Indonesia*, Sustainable Development Goals Series, https://doi.org/10.1007/978-981-96-4182-6_8

after the disaster and pays less attention to the pre-disaster stage (Supratiwi et al. 2022).

Since the 1980s, various international declarations have promulgated a disaster risk reduction (DRR) and resilience paradigm as the guiding basis of disaster management and development agencies in all countries (Imperiale and Vanclay 2020; UNISDR 2007; UNITED NATIONS 2015). According to the United Nations, resilience is defined as the ability of a system, community, or society exposed to hazards to resist, absorb, accommodate, adapt to, transform, and recover from the effects of a hazard in a timely and efficient manner, including through the preservation and restoration of its essential basic structures and functions through risk management (UNDRR 2019).

Disaster studies offer a lens through which to look deeply at society and its underlying social behaviors (Reinhardt and Ross 2019). Natural hazard-induced disasters are one of the severe challenges to the sustainability of communities around the world. The occurrence of earthquakes, floods, storms, and various other disasters can have a destructive impact, threaten the safety of life, and harm the basis of people's lives. In the face of the complexity and intensity of disaster risk, community resilience emerged as a significant paradigm in Disaster Risk Reduction (DRR) efforts. This chapter aims to delve deeper into community resilience's role in disaster risk reduction by analyzing key concepts such as social cohesion, adaptive capacity, and community-based strategies.

Through the lens of social sciences, this chapter tries to provide a comprehensive understanding of the dynamics of community resilience. In addition, we will highlight the context-specific nature of community resilience, showing that a one-size-fits-all approach is not suitable for all. The integration of community resilience into DRR policies and practices is a key focus in efforts to reduce vulnerability and increase community capacity in dealing with disasters.

8.2 Concept of Community Resilience

8.2.1 Definition of Community Resilience

Community resilience is a concept that can be seen through the lens of social science (Ómarsdóttir et al. 2022). Community resilience is the ability of a community to anticipate, prepare, adapt, and recover from disruptions, stressors, and trauma through ongoing planning and technical solutions (Kramer et al. 2023; Sousa and Moss 2022). The system can manage shocks and return to a stable state in response to extreme events (Jamal and Hasan 2023). Community resilience is the process that links networks of adaptive capacity with adaptation after disruption or adversity. This adaptive capacity includes economic development, social capital, information and communication, and community competence (Kulig et al. 2008; Nguyen and Akerkar 2020). Factors that contribute to community resilience include physical infrastructure, disaster conditions, and socioeconomic characteristics of a community (Kendra et al. 2021; Xu et al. 2022).

Community resilience is important in controlling and responding to disasters, such as hurricanes, tsunamis, floods, and pandemics. Community resilience involves strengthening governance, improving public relations, and improving risk-coping styles (Aldrich et al. 2015). Community resilience is a dynamic process involving social and monetary capital, technological expertise, physical and organizational infrastructure, shared philosophies, survival patterns and cultures, emotional qualities, and norms around cooperation and interdependence (Sousa and Moss 2022; Suleimany et al. 2022). Developing models and practices around community resilience is critical to coping with and recovering from collective stress and trauma.

8.2.2 Factors Influencing Community Resilience

Community resilience is characterized by interdependence, interaction, and the ability to overcome challenges. It emphasizes the importance of community collaboration, shared identity, and empowerment in promoting resilience (Murray and Zautra 2012). It is influenced by physical and social infrastructure, population characteristics, conceptual characteristics, and problem-solving processes (Haggard et al. 2019). Various factors influence community resilience. These include social and monetary capital, technological expertise, strong physical and organizational infrastructure, shared philosophies, patterns and cultures of survival and meaning-making, emotional qualities such as optimism and trust, and norms around cooperation and interdependence (Norris et al. 2008; Sousa and Moss 2022), and a sense of belonging in the community (Quinn et al. 2021). Other factors that contribute to community resilience are age, history of depression or anxiety, place of residence after a disaster, and receiving support from the government (Obuobi-Donkor et al. 2022).

Additionally, factors such as gender, marital status, profession, and employment also play a role in determining resilience levels (Mohan and Deb 2023). The income level, vulnerability of the population, and the built environment are identified as key factors affecting community resilience during global pandemics like COVID-19 (Xu et al. 2020). Furthermore, knowledge assets, SECI (Socialization, Externalization, Combination, and Internalization), enablers, and disaster preparedness are important factors in increasing community resilience (Oktari et al. 2022). Haase et al. (2021) show that there are six capacities of community resilience, namely social, economic, physical, human, institutional, and environmental.

Meanwhile, Robertson et al. (2021) identify those that contribute to community resilience, including social bonding, leadership, integration, communication, and training. Community resilience, the ability of societies to withstand and recover from stressful events, has gained increased attention in recent years. In this context, one of the key concepts is social capital, which includes features like collective action, social embeddedness, organizational linkages, citizen participation, a sense of community, and social support (Aldrich 2011; Aldrich 2012; Norris et al. 2008; Ómarsdóttir et al. 2022).

For developing countries, community resilience is more complex because the human system is more vulnerable and massive development activities. On the other hand, the development process can generate new hazardous areas. Development is a series of human activities to achieve a common goal, such as transportation systems, communication systems, public services, including building and housing, and economic development. Development in disaster-prone areas is generally more complex because poor planning will produce more losses than benefits. There is much debate about the influence of development or human activities relative to disaster, particularly in developing countries. However, the impact of the disaster was found to be greater in developing countries, and it related to the historical development of these countries (Alcántara-Ayala 2002).

Nowadays, natural hazard-induced disasters in Indonesia are also identified due to ill-planned development (Marfai et al. 2008). For instance, coastal inundation due to land subsidence is one of the most severe problems faced by coastal cities like Jakarta and Semarang. The use of the coastal and low-lying areas for economic activities, housing, and other human activities contributes to extensive land use change and increasing groundwater extraction, leading to land subsidence. The subsidence has damaged the road and dwelling units, overflowing water into residential areas, permanently inundation of the coastal areas, and change in the coastline. River flooding is also one of the most frequently occurring disasters in Indonesia. Land use changes from forest areas to intensive agricultural land and tourist spots have contributed to increasing surface runoff and degradation of watersheds, leading to flooding in the lowland areas. It is very clear from the case of the Ciliwung flood in Jakarta due to land use change in the upstream areas (i.e., Bogor).

Mountain regions have natural and scenic values that make them attractive for leisure and tourism activities. So, land occupation and population in many mountain areas are increasing. Altitude and climate in mountain areas make geomorphic processes active and sensitive to environmental change (e.g., climate and land use). Among such processes, slope movements get particular attention because they are severe hazards for human life and property. The slope movements occur due to a combination of various parameters triggered by rainfalls, earthquakes, or human activity (Sterlacchini et al. 2011). Gorsevski et al. (2006) describe human activities that can generate slope movements, including urban expansion, road-building, and deforestation. Regmi et al. (2010) underline that road-cuts are usually sites of anthropologically induced instability of slopes. The number of occurrences of slope movements or landslides in Indonesia has been increasing for the last ten years. Hadmoko et al. (2010) reported that landslides in Indonesia are triggered by heavy rainfall and human activity. Development is expanding into unstable areas causing slope instability problems.

People in disaster-prone zones are generally marginal or poor due to low education levels and weak economies. Both cause them to lose competition in employment and economically can only afford to buy and live in non-strategic and disaster-prone areas. Poverty and education are like "vicious chains," difficult to break unless there are internal and external energies that unravel them. Individuals or communities living in poverty have limited access to economic, educational, and health resources. This can reduce their capacity to respond or prepare for disasters. Poor communities have poor infrastructure and limited access to equipment and training facilities for disaster relief. This inability can be detrimental to efforts to build resilience to disasters. In addition, poor individuals/communities are faced with poor environmental sanitation, so they are less healthy but limited in obtaining health services. This condition increases vulnerability to the impact of disasters such as infectious diseases or malnutrition. Community awareness and education are key elements in improving disaster resilience. Poverty can hamper community outreach and education efforts related to disasters.

8.2.3 The Relationship Between Resilience and Disaster Risk Reduction

Disasters are caused by a combination of environmental and sociocultural factors (Tierney 1999), and it is important to recognize the social vulnerability of communities, which is not evenly distributed among social groups or between places. Some communities may be more susceptible to the impacts of hazards than others due to their location or population-level characteristics such as race, income, and age (Cutter and Emrich 2006; Cutter and Finch 2008). The social vulnerability of a population, defined as the capacity of a social group to anticipate, cope with, resist, and recover from an incident, determines the community's resilience no less than the concrete, physical components of disaster preparedness and response (Wisner et al. 2004). Communities with strong social ties that connect residents and enable reciprocity and collective action are better able to adapt and recover from the impact of disasters (Szreter 2004).

Resilience and disaster risk reduction are closely related concepts. Resilience refers to the ability of a system or community to withstand and recover from the impacts of disasters. In contrast, disaster risk reduction focuses on reducing vulnerabilities and exposure to hazards. The Sendai Framework for Disaster Risk Reduction emphasizes the need for inclusive and multi-hazard approaches to DRR, as well as the synergies with climate change, health, and sustainable development (Trut and Kovačević 2022). Climate change is already increasing the frequency and intensity of natural hazards, making resilience building and DRR even more important (Mal et al. 2018). Governments that integrate risk management into their governance frameworks have a better record of DRR and resilience building (Yasukawa and Delgado 2023). DRR and climate change adaptation share a common area of interest and partly overlap, highlighting the need to consider the impact of climate change on disaster risks (Forino et al. 2015). Resilience building and DRR are essential for reducing vulnerabilities and improving preparedness at all levels (Shaw et al. 2022).

A lesson can be seen from the landslide disaster in Ciweday, West Java, in 2010. The Ciweday landslide occurred in the tea plantation and tea factory area of Tenjolaya Bandung Village, which caused damage to tea land, factory buildings, and workers' housing and the death of a number of tea plantation workers. Damage to land and factories will affect the supply chain of raw materials and will greatly harm companies, so companies generally have disaster risk management. The area affected by the avalanche and its communities are still the tea company's responsibility. They have received some disaster training and have access to health. So that communities affected by the Ciweday landslide recover faster and return to the garden to maintain the tea supply chain. Disaster risk reduction (DDR) is carried out through disaster drills, early warning system development, and disaster management systems.

8.3 Theoretical Foundation and Conceptual Framework

8.3.1 Social Cohesion in the Context of Community Resilience

Social cohesion plays a crucial role in community resilience (Ludin et al. 2019; Svyrydenko et al. 2022). It is considered the glue that holds society together and is essential for coping and recovery in the face of collective stressors and trauma (Saghin et al. 2022). The COVID-19 pandemic and the lockdown have highlighted the importance of involving the population in solving problems that affect the existence and trajectory of individuals and civil society (Cahyono et al. 2023; Gongora-Svartzman and Ramirez-Marquez 2022). Crisis conditions, such as war and the pandemic, have an impact on social cohesion, with some aspects increasing and others decreasing (Svyrydenko et al. 2022). Social media communications and communities can be used to measure and understand social cohesion during disasters, providing valuable information for risk management and disaster mitigation policies (O'Brien 2022). The resilience of a community depends not only on tangible factors like social and monetary capital, but also on intangible factors like shared philosophies, emotional qualities, and norms around cooperation and interdependence. Overall, social cohesion is a key component of community resilience, and understanding its dynamics can help in developing strategies to enhance resilience in the face of crises.

8.3.2 Adaptive Capacity and Community Adjustment

Adaptive capacity is crucial for community resilience and the ability to adjust and respond to change. It involves the personal attributes possessed by community residents that enable them to survive and recover during and after a disaster (Landau et al. 2021). Strong civic engagement, social networks, and community-based organizations contribute to enhanced adaptive capacity (Chukwuemeka Ogugua 2022). Access to assets, diversity and flexibility, learning and knowledge, and natural capital are important domains that influence adaptive responses at the community level (Green et al. 2021). At the household level, adaptive responses are supported by strong governance or institutions and can substitute learning knowledge and natural capital with each other (Jacobs et al. 2015). Measuring adaptive capacity and transformative capacity is essential for understanding community vulnerability and resilience, but operational definitions vary widely (Berman et al. 2017). Building adaptive capacity is essential for managing natural resources and meeting future changes with confidence.

Environmental change impacts people, both individually and collectively, with the most visible impacts often manifesting in personal losses, including the loss of lives, livelihoods, capital, mental well-being, and physical health (Steffen et al. 2018). Thus, much of the adaptive capacity scholarship in the environmental social sciences has focused on analyses of individuals and households (Siders 2019) without explicit connection to the scalar impacts of the environmental change processes under investigation. This is problematic because many of the most pressing environmental changes today—increased severity and duration of drought, regional vegetation shifts, and sea-level rise, among others—impact landscapes differently across space and time, but also across myriad human-defined jurisdictions of local and regional governance, including the boundaries of local communities and the borders of sovereign nations (Cumming et al. 2014; Haider et al. 2018).

The capacities necessary for humans to adapt to such manifestations of environmental change are functions of individual and societal characteristics and often a function of nested capacities across communities and jurisdictions (Adger et al. 2009; Hill and Engle 2013). Thus, it is critical to evaluate adaptive capacities at levels beyond just individuals and households; levels at which elements such as decision-making authority, institutional structures, infrastructure, and/or coordination or collaborative capacities increase in importance (Allen and Holling 2010) but are often understudied.

8.3.3 Community-Based Strategies for Disaster Risk Reduction

Community-based strategies for disaster risk reduction (CB DRR) are practical approaches for emergency management that focus on addressing the specific needs and context of communities (Maher and Toh 2023). These strategies involve consultation and coordination among policymakers, civil societies, communities, and other stakeholders (Roy et al. 2023). Quantitative assessments have shown that CB DRR projects can be effective in reducing disaster losses, with regional differences influenced by economic level and hazard severity (Wu et al. 2023). Implementing CB DRR programs can improve community knowledge and skills in disaster mitigation (Van Niekerk et al. 2018). However, there is a need for greater community participation, better coordination among stakeholders, and increased education and socialization on disaster management to enhance the effectiveness of CB DRR interventions (Bali 2022; Hamzah et al. 2022; Suryani et al. 2022). Community empowerment and cross-sector collaboration involving the community, government, and universities are crucial for successful CB DRR initiatives.

Community-based strategies in disaster risk reduction face both challenges and opportunities. One of the challenges is the complex network of community, technology, and user motivation, which makes it difficult to encourage users to adopt preparation efforts instead of reactive measures (Maher and Toh 2023). Another challenge is the lack of consideration of local people's needs and voices, which undermines the success of various disaster risk reduction intervention programs (Roy et al. 2023). However, there are also opportunities for implementing community-based strategies. Community participation and coordination among stakeholders are crucial for effective and successful disaster risk reduction interventions (Paschalia et al. 2022). Community-based disaster risk reduction programs can improve the knowledge and skills of the community in disaster mitigation, leading to better preparedness and response (Willison et al. 2022). Involving communities in disaster risk reduction can also help reduce the burden on government and local administration and make communities more self-reliant in emergencies (Bali 2022). Capacity building and awareness creation among communities are important for building resilience to disasters.

Indonesia, as a disaster-prone country, demands the active role of the community in disaster management. So, the perspective that focuses on handling during and after a developing disaster must be changed. In this matter, the government established the Desa Tangguh Bencana (Destana)/Disaster Resilient Village Policy. With various activities with community involvement, the risk and impact of disasters can be eliminated because the community has been trained to prepare for disasters (Supratiwi et al. 2022).

8.4 Case Studies

8.4.1 Community Resilience in Facing Natural Hazard-Induced Disasters

Community resilience in facing natural hazard-induced disasters is a critical aspect of disaster risk reduction and sustainable development. Research has shown that community resilience is influenced by various factors such as age, self-efficacy, well-being, community leadership, and preparedness (Zaman and Raihan 2023). Building resilient communities requires a comprehensive multiphase research model that considers the socio-ecological context and includes major theoretical dimensions (Zander et al. 2023). Community resilience plays a significant role in the relationship between tourism and quality of life (QOL) during disasters and crises (Yang and Kim 2023).

Transformative resilience is crucial in the building process of community resilience, which involves the transformation phases, heterogeneity and interdependence of capacities, and the intensity of resources used (Crews et al. 2023). Educating the public and stakeholders about natural hazards is essential for fostering community resilience (Maurischa et al. 2023). Overall, understanding the current trends in research methods, dimensions, and theoretical standpoints is crucial for effective policymaking and supporting resilient and sustainable development in communities facing natural hazard-induced disasters.

In Banda Aceh and Aceh Besar, Indonesia, early recovery income support, physical and mental health, ability to transfer to other jobs/skills, availability and timeliness of livelihood support, together with its cultural sensitivity and governance structure, are among the most important factors. Given the nature of resettlement, access to infrastructure, location of relocated sites, and neighborhood safety also played an important role in establishing sustained employment for these relocated communities in Indonesia (Sina et al. 2019).

8.4.2 Impact of Community Resilience in Enhancing Disaster Response

Community resilience plays a crucial role in enhancing disaster response (Jamal and Hasan 2023; Lim et al. 2022). Communities need to be able to manage shocks and return to a steady state after an extreme event (Jo and Lee 2022). Studies have shown that community resilience can mitigate the negative impacts of disaster damage on the recovery level of survivors (Willar et al. 2022).

Factors such as disruptions in physical infrastructures, disaster conditions, and socioeconomic characteristics of a community can affect the level of community resilience. Understanding the factors associated with the loss of community resilience can help improve hazard preparedness, enhance disaster management practices, and strengthen infrastructure and community resilience. By enhancing community resilience, communities can better prepare for, respond to, and recover from disasters, ultimately improving their overall disaster response.

In The Merapi Volcano Community, Java, Local communities generally play a crucial role during a disaster, so their involvement in pre-disaster capacity development may prove beneficial in the face of a disaster threat. Thus, People-Centered Disaster Risk Reduction (PCDRR) programs could enable communities living in disaster-prone areas to become more resilient (Mutiarni et al. 2022). In Simeulue (Sutton et al. 2022), the community has the resilience to face tsunamis and responds well.

8.4.3 Integration of Community Resilience in Disaster Risk Reduction Policies

Disaster risk reduction (DRR) policies should integrate community resilience to manage climate-related disasters effectively and reduce their impacts. The net benefits of DRR, including economic, ecological, and social co-benefits, need to be considered to strengthen the case for DRR as part of sustainable development (Rözer et al. 2023). However, there is a disconnect between available planning tools and the evidence of materialized multiple resilience dividends at the community level, hindering successful implementation (Roy et al. 2023). Greater community participation, consultation, and coordination among stakeholders are crucial for effective DRR interventions (Willison et al. 2022).

Community resilience involves improving local people's awareness, capacities, and support to adapt to climate change and prevent hazards from becoming disasters (Gil-Rivas and Kilmer 2016). Vulnerability and risk factors at the community and household levels should be assessed, and community-based DRR approaches should be implemented to develop resilience strategies (Van Niekerk et al. 2018). Inclusive planning efforts considering marginalized community members' perspectives and broadening the definition of critical facilities are necessary to reduce vulnerability and increase adaptive capacity. Imperiale and Vanclay (2021) conclude that a shift from centralized civil protection to decentralized, inclusive community empowerment systems is needed to make the disaster risk reduction and resilience paradigm more effective.

8.5 Discussion

8.5.1 The Role of Community Resilience in Disaster Risk Reduction

Community resilience plays a crucial role in disaster risk reduction (Mutiarni et al. 2022). Understanding how well communities are prepared for natural hazards and how well they can cope with and recover from shocks is essential for effective disaster management (Zander et al. 2023). Investing in disaster risk reduction interventions can reduce potential impacts from disasters and deliver broader economic, ecological, and social co-benefits (Rözer et al. 2023). Women's groups in rural communities have been found to provide financial, social, socio-political, and public health support, contributing to building safer, sustainable, and resilient communities (Khatri et al. 2023). Community-based disaster risk reduction approaches focus on identifying community-specific needs and context to develop technologies that help mitigate the effects of disasters (Maher and Toh 2023). By understanding the factors contributing to community resilience, such as self-efficacy, well-being, and preparedness, efforts can be prioritized to support communities with low capacity to prepare and recover from natural hazards.

Disaster community resilience in Indonesia is an important issue due to the high risk of natural hazard-induced disasters in Indonesia. Previous research has highlighted the need for local knowledge, awareness, and wisdom to improve social systems and create disaster resilience (Agustianingsih et al. 2023; Cahyono et al. 2021). The Indonesian government has allocated responsibility for disaster governance mechanisms to local governments. However, the regulatory framework needs to be reviewed to address issues such as determining disaster status, budget allocation, and strengthening the capacity of vulnerable communities (Firdaus et al. 2023). Community preparedness for disasters is critical, and there is a need for community-based disaster management training to improve knowledge, attitudes, and skills (Prianto and Abdillah 2023). A comprehensive approach that integrates learning, local wisdom, and community-based initiatives is needed to improve disaster community resilience in Indonesia.

8.5.2 Challenges and Barriers to Enhancing Community Resilience

Challenges and barriers to enhancing community resilience include the following: a lack of understanding of community resilience and the social dimensions of risk, which limits the implementation of effective disaster risk reduction (DRR) and resilience-building strategies (Cox 2023); a paternalistic social protection culture and a command-and-control approach to knowledge and resources for risk reduction, which hinder the development of a global culture of community well-being and resilience (Virapongse et al. 2022); the centralization of knowledge and resources by civil protection authorities, which stifles the capacities of local communities to reduce disaster risks and impacts (Imperiale and Vanclay 2021); inequity in the scientific process, gaps in data ethics and governance, a mismatch of scale and focus, and a lack of actionable information for communities when applying Earth science data to resilience efforts (Imperiale and Vanclay 2020); and communities becoming politically balkanized and antagonistic toward sustainability- or resilience-oriented community plans, which creates barriers to accomplishing sustainable and resilient planning (Buckman 2022), funding, the difficulty of including local knowledge in resilience building (Berke et al. 2021).

8.5.3 Enhancing the Integration of Community Resilience in Disaster Risk Reduction

Enhancing the integration of community resilience in disaster risk reduction requires a structured consideration of multiple resilience dividends from planning to monitoring and evaluation stages (Rözer et al. 2023). This approach is crucial to secure local buy-in and ensure these dividends materialize as intended (Zander et al. 2023). However, there is a disconnect between available planning tools and the evidence of materialized multiple resilience dividends, hindering successful community application (Roy et al. 2023). Greater community participation and better coordination among stakeholders are urgently needed to effectively prioritize community needs and voices in disaster risk reduction interventions (De Risi et al. 2022). Additionally, community resilience plays a moderating role in the relationship between disaster damage types and the recovery level of survivors (Lim et al. 2022). Therefore, efforts should be made to enhance community resilience through social welfare practices, as it is crucial for increasing survivors' recovery level.

A holistic approach is crucial in disaster risk reduction as it encompasses a comprehensive and integrated system that deals with all sorts of emergencies, risks, and crises. It involves multiple sectors and stakeholders and requires a combination of physical and social measures, with full participation of affected populations and other stakeholders (Hertelendy et al. 2024; Tong and Zhang 2020). This approach allows for a multi-disciplinary perspective, considering various factors such as vulnerability, exposure to hazards, and the complex interactions among social, economic, political, and physical factors (Ishiwatari and Sasaki 2023). By adopting a holistic approach, organizations can better address the increasing frequency, duration, and intensity of crisis events and focus on the top hazards they may face (Nguyen et al. 2022). It also enables the development of interdisciplinary approaches that can resolve a wide range of issues, including risk and damage assessment, behavior change, investment, and evidence-based policy formulation (Alexander 2020). A holistic approach to disaster risk reduction is essential for effective preparedness, response, and recovery efforts.

Natural hazard-induced disasters occur throughout the year, prompting local communities and the government to adopt various protective measures. These measures are broadly categorized into structural and non-structural approaches. Structural measures include infrastructure such as dyke systems and floodwalls to manage river flooding, early warning systems for earthquakes, tsunamis, and landslides, retaining walls to stabilize slopes, and facilities like dams, spillways, control gates, and flood detention basins. Non-structural measures focus on activities such as land use planning and management, public education and training programs, and reforestation efforts in coastal, upstream, and mountainous areas. As a developing nation, Indonesia has emphasized non-structural mitigation strategies, as they require fewer financial resources than large-scale infrastructure projects. However, disaster risks are exacerbated by low public awareness of natural hazards. Consequently, disaster management efforts in Indonesia today prioritize raising community awareness and fostering proactive engagement in risk reduction initiatives.

The government, academic institutions, and the public collaborate on disaster management, with a strong emphasis on raising awareness. However, fostering community awareness can be complex due to the complexities of human systems. As a result, disaster management efforts prioritize educating and informing communities about hazards and self-rescue measures through various media channels, including newspapers, television, radio, and posters. These efforts involve the active participation of the government, academic institutions, and NGOs. A community empowerment approach is adopted to enhance awareness, as it encourages active participation in every phase of disaster management.

8.6 Conclusion

The concept of community resilience can be understood through the lens of social science to provide a comprehensive understanding of its significance in mitigating the impact of disasters. Key concepts such as social cohesion, adaptive capacity, and community-based strategies with greater emphasize on the dynamic and context-specific nature of community resilience. It is important to consider the integration of community resilience into Disaster Risk Reduction (DRR) policies and practices. Quite a few communities have leveraged their resilience to reduce their vulnerability and increase their capacity to respond effectively to disasters.

Building resilient communities to disaster risk reduction requires an integrated assessment of the existing condition and developing a comprehensive action plan to motivate and control collective effort in different systems of the cities to mitigate probable damages and casualties, absorb shocks, and facilitate the redevelopment processes. Indeed, efforts to create resilient communities are not limited to before or during the outbreak of the pandemic, and this needs to be a continuous process.

A holistic approach is needed in disaster risk reduction by emphasizing the vital role of community resilience. It provides valuable insights for policymakers, practitioners, and researchers working in DRR by offering practical recommendations to increase community resilience and build a more secure and sustainable future amid the growing threat of disasters.

References

Adger WN, Eakin H, Winkels A (2009) Nested and teleconnected vulnerabilities to environmental change. Front Ecol Environ 7(3):150–157. https://doi.org/10.1890/070148

Agustianingsih DP, Ariyaningsih, Shaw R (2023) Community disaster resilience using multi-hazard assessment during Covid-19: the case of Denpasar, Indonesia. Nat Hazards Res 3(3):572–582. https://doi.org/10.1016/j.nhres.2023.04.006

Alcántara-Ayala I (2002) Geomorphology, natural hazards, vulnerability and prevention of natural hazard induced-disasters in developing countries. Geomorphology 47(2–4):107–124. https://doi.org/10.1016/S0169-555X(02)00083-1

Aldrich DP (2011) The power of people: social capital's role in recovery from the 1995 Kobe earthquake. Nat Hazards 56(3):595–611. https://doi.org/10.1007/s11069-010-9577-7

Aldrich DP (2012) Building resilience: social capital in post-disaster recovery. University of Chicago Press, Chicago

Aldrich DP, Sawada Y, Oum S (2015) Approaches towards effective disaster risk-coping strategies and regional cooperation on disaster management. In: Resilience and recovery in Asian disasters. Springer, London, pp 339–353. https://doi.org/10.1007/978-4-431-55022-8_16

Alexander D (2020) Disaster risk reduction. In: Oxford research encyclopedia of politics. Oxford University Press, Oxford. https://doi.org/10.1093/acrefore/9780190228637.013.1538

Allen CR, Holling CS (2010) Novelty, adaptive capacity, and resilience. Ecol Soc 15(3):art24–art24. https://doi.org/10.5751/ES-03720-150324

Bali R (2022) Importance of community awareness and preparedness in disaster risk reduction. Res Rev Int J Multidiscip 7(10):40–57. https://doi.org/10.31305/rrijm.2022.v07.i10.005

Berke PR, Quiring SM, Olivera F, Horney JA (2021) Addressing challenges to building resilience through interdisciplinary research and engagement. Risk Anal 41(7):1248–1253. https://doi.org/10.1111/risa.13202

Berman M, Kofinas G, BurnSilver S (2017) Measuring community adaptive and transformative capacity in the arctic context, pp 59–75. https://doi.org/10.1007/978-3-319-46150-2_6

Buckman S (2022) Resilience planning in antagonistic communities. In: Investing in disaster risk reduction for resilience. Elsevier, London, pp 123–140. https://doi.org/10.1016/B978-0-12-818639-8.00007-7

Cahyono SA, Wuryanta A, Lastiantoro CY (2021) The local knowledge to mitigate the landslide disaster in Beruk village, Jatiyoso sub-district, Karanganyar regency. IOP Conf Ser Earth Environ Sci 874(1):012015–012015. https://doi.org/10.1088/1755-1315/874/1/012015

Cahyono SA, Agus C, Raharjo SAS, Nugroho AW, Indrajaya Y (2023) Strategic tropical natural resources management and household socioeconomic resilience due to the Covid-19 pandemic. In: Agricultural bioeconomy. Elsevier, London, pp 167–182. https://doi.org/10.1016/B978-0-323-90569-5.00007-X

Chukwuemeka Ogugua A (2022) Community resilience and adaptive capacity: an integrative perspective. Int J Commun Resilience. https://doi.org/10.51595/11111162

Cox LA (2023) Decision theory challenges for catastrophic risks and community resilience, pp 157–183. https://doi.org/10.1007/978-3-031-32013-2_5

Crews T, Goodwin C, Rowe S, Olmeta-Schult F (2023) The quest to support community resilience in the face of coastal hazards and natural hazard induced-disasters. Curr J Mar Educ 37(2):3–7. https://doi.org/10.5334/cjme.81

Cumming GS, Buerkert A, Hoffmann EM, Schlecht E, von Cramon-Taubadel S, Tscharntke T (2014) Implications of agricultural transitions and urbanization for ecosystem services. Nature 515(7525):50–57. https://doi.org/10.1038/nature13945

Cutter S, Emrich C (2006) Moral Hazard, Social catastrophe: the changing face of vulnerability along the hurricane coasts. Annal AAPSS 604(2006):102–112

Cutter S, Finch C (2008) Temporal and spatial changes in social vulnerability to natural hazards damage. Proc Natl Acad Sci USA 2008:1

De Risi R, Muhammad A, De Luca F, Goda K, Mori N (2022) Dynamic risk framework for cascading compounding climate-geological hazards: a perspective on coastal communities in subduction zones. Front Earth Sci. https://doi.org/10.3389/feart.2022.1023018

Firdaus A, Lestari F, Afiff SA, Herdiansyah H (2023) Integration of knowledge and local wisdom for disaster resilience in Anak Krakatau volcano. Jàmbá J Dis Risk Stud. https://doi.org/10.4102/jamba.v15i1.1457

Forino G, von Meding J, Brewer GJ (2015) A conceptual governance framework for climate change adaptation and disaster risk reduction integration. Int J Dis Risk Sci 6:372–384

Gil-Rivas V, Kilmer RP (2016) Building community capacity and fostering disaster resilience. J Clin Psychol 72(12):1318–1332

Gongora-Svartzman G, Ramirez-Marquez JE (2022) Social cohesion: mitigating societal risk in case studies of digital media in hurricanes Harvey, Irma, and Maria. Risk Anal 42(8):1686–1703. https://doi.org/10.1111/risa.13820

Gorsevski PV, Gessler PE, Boll J, Elliot WJ, Foltz RB (2006) Spatially and temporally distributed modeling of landslide susceptibility. Geomorphology 80(3–4):178–198. https://doi.org/10.1016/j.geomorph.2006.02.011

Green KM, Selgrath JC, Frawley TH, Oestreich WK, Mansfield EJ, Urteaga J, Swanson SS, Santana FN, Green SJ, Naggea J, Crowder LB (2021) How adaptive capacity shapes the Adapt, React, Cope response to climate impacts: insights from small-scale fisheries. Clim Change 164(1–2):15–15. https://doi.org/10.1007/s10584-021-02965-w

Haase TW, Wang W-J, Ross AD (2021) The six capacities of community resilience: evidence from three small Texas communities impacted by Hurricane Harvey. Nat Hazards 109(1):1097–1118. https://doi.org/10.1007/s11069-021-04870-y

Hadmoko DS, Lavigne F, Sartohadi J, Hadi P, Winaryo (2010) Landslide hazard and risk assessment and their application in risk management and landuse planning in eastern flank of Menoreh Mountains, Yogyakarta Province, Indonesia. Nat Hazards 54(3):623–642. https://doi.org/10.1007/s11069-009-9490-0

Haggard R, Cafer A, Green J (2019) Community resilience: a meta-study of international development rhetoric in emerging economies. Community Dev 50(2):160–180. https://doi.org/10.1080/15575330.2019.1574851

Haider LJ, Boonstra WJ, Peterson GD, Schlüter M (2018) Traps and sustainable development in rural areas: a review. World Dev 101:311–321. https://doi.org/10.1016/j.worlddev.2017.05.038

Hamzah A, Sofyana H, Cahyaningsih H, Hufad A, Hasanah VR, Handayani D, Mulyo GPE (2022) Community-based information for disaster risk identification. Dis Emerg Med J 7(4):209–214. https://doi.org/10.5603/DEMJ.a2022.0032

Hertelendy AJ, Jaiswal R, Donahue J, Reilly MJ (2024) Disaster risk management. In: Ciottone's disaster medicine. Elsevier, London, pp 178–190. https://doi.org/10.1016/B978-0-323-80932-0.00030-6

Hill M, Engle NL (2013) Adaptive capacity: tensions across scales. Environ Policy Gov 23(3):177–192. https://doi.org/10.1002/eet.1610

Imperiale AJ, Vanclay F (2020) Barriers to enhancing disaster risk reduction and community resilience: evidence from the L'Aquila disaster. Polit Gov 8(4):232–243. https://doi.org/10.17645/pag.v8i4.3179

Imperiale AJ, Vanclay F (2021) Conceptualizing community resilience and the social dimensions of risk to overcome barriers to disaster risk reduction and sustainable development. Sustain Dev 29(5):891–905. https://doi.org/10.1002/sd.2182

Ishiwatari M, Sasaki D (2023) Special issue "Disaster risk reduction and climate change adaptation: an interdisciplinary approach". Int J Environ Res Public

Health 20(3):2641–2641. https://doi.org/10.3390/ijerph20032641
Jacobs B, Nelson R, Kuruppu N, Leith P (2015) An adaptive capacity guide book: assessing, building and evaluating the capacity of communities to adapt in a changing climate. University of Technology Sydney and University of Tasmania
Jamal TB, Hasan S (2023) Understanding the loss in community resilience due to hurricanes using Facebook data. Int J Dis Risk Reduct 97:104036–104036. https://doi.org/10.1016/j.ijdrr.2023.104036
Jo YS, Lee JE (2022) Impact of the core system on local community disaster resilience. Crisis Emerg Manag Theory Praxis 18(11):61–78. https://doi.org/10.14251/crisisonomy.2022.18.11.61
Kendra J, Clay L, Gill K, Trivedi J, Marlowe V, Aguirre B, Nigg J, Trainor J, Carbone E, Links J (2021) Community resilience: toward a framework for an integrated, interdisciplinary model of disaster. Nat Hazards Rev. https://doi.org/10.1061/(ASCE)NH.1527-6996.0000495
Khatri J, Tippett V, Durham J (2023) The role of self-help women's groups in disaster risk reduction and community resilience in Nepal. Prehosp Disaster Med 38(S1):s71–s72. https://doi.org/10.1017/S1049023X23002121
Kramer A, Belding S, Coney K (2023) Community resilience options: a menu for enhancing local energy resilience
Kulig JC, Edge DS, Joyce B (2008) Understanding community resiliency in rural communities through multi-method research. J Rural Community Develop 3(3):1
Landau LF, Campbell LK, Svendsen ES, Johnson ML (2021) Building adaptive capacity through civic environmental stewardship: responding to COVID-19 alongside compounding and concurrent crises. Front Sustain Cities. https://doi.org/10.3389/frsc.2021.705178
Lim HS, Nam DY, Lee KY (2022) The effect of community resilience on the recovery level of disaster victims. Center Soc Welf Res Yonsei Univ 75:5–30. https://doi.org/10.17997/SWRY.75.1.1
Ludin SM, Rohaizat M, Arbon P (2019) The association between social cohesion and community disaster resilience: a cross-sectional study. Health Soc Care Community 27(3):621–631. https://doi.org/10.1111/hsc.12674
Maher T, Toh C (2023) We always think it's never going to happen to us: understanding what motivates communities to engage in emergency preparedness. Proc Design Soc 3:19–28. https://doi.org/10.1017/pds.2023.3
Mal S, Singh R, Huggel C, Grover A (2018) Introducing linkages between climate change, extreme events, and disaster risk reduction. In: Climate change, extreme events and disaster risk reduction: towards sustainable development goals, pp 1–14
Marfai MA, King L, Sartohadi J, Sudrajat S, Budiani SR, Yulianto F (2008) The impact of tidal flooding on a coastal community in Semarang, Indonesia. Environmentalist 28(3):237–248. https://doi.org/10.1007/s10669-007-9134-4
Maurischa SD, Fahmi FZ, Suroso DSA (2023) Transformative resilience: transformation, resilience and capacity of coastal communities in facing disasters in two Indonesian villages. Int J Dis Risk Reduct 88:103615–103615. https://doi.org/10.1016/j.ijdrr.2023.103615
Mohan M, Deb R (2023) Factors associated with resilience among people with spinal cord injury in the community setting. Iran Rehab J 21(1):157–166. https://doi.org/10.32598/irj.21.1.1808.1
Murray K, Zautra A (2012) Community resilience: fostering recovery, sustainability, and growth. In: The social ecology of resilience. Springer, New York, pp 337–345. https://doi.org/10.1007/978-1-4614-0586-3_26
Mutiarni YS, Nakamura H, Bhattacharya Y (2022) The resilient community: strengthening people-centered disaster risk reduction in the Merapi Volcano Community, Java, Indonesia. Sustainability 14(4):2215–2215. https://doi.org/10.3390/su14042215
Nguyen HL, Akerkar R (2020) Modelling, measuring, and visualising community resilience: a systematic review. Sustainability 12(19):7896–7896. https://doi.org/10.3390/su12197896
Nguyen P, Tran N, Shamshirband S, Tran T, Quan T (2022, 2022/10) Comprehensible alarm message generation for natural disaster. In: 2022 9th NAFOSTED conference on information and computer science (NICS)
Norris FH, Stevens SP, Pfefferbaum B, Wyche KF, Pfefferbaum RL (2008) Community resilience as a metaphor, theory, set of capacities, and strategy for disaster readiness. Am J Community Psychol 41(1–2):127–150. https://doi.org/10.1007/s10464-007-9156-6
Nuryanti L, Sulivan G, Wang S, Branney P (2017) Resilience as boundary object in disaster policy, research and interventions: border thinking from the indonesian supermarket of disaster. In: Sullivan GBCJEBMMSE (ed) Resistance and renewal in theoretical psychology, vol 1. Captus Press, London, pp 163–172
O'Brien K (2022) Corrigendum to social cohesion and resilience in first Australian family and kinship networks. J Fam Hist 47(3):361–362. https://doi.org/10.1177/03631990221087792
Obuobi-Donkor G, Eboreime E, Shalaby R, Agyapong B, Adu MK, Owusu E, Mao W, Oluwasina F, Pazderka H, Agyapong VIO (2022) Evaluating community resilience and associated factors one year after the catastrophic fort Mcmurray flood. Int J Environ Res Public Health 19(23):16153–16153. https://doi.org/10.3390/ijerph192316153
Oktari RS, Latuamury B, Idroes R, Sofyan H, Munadi K (2022) Validating knowledge creation factors for community resilience to disaster using structural equation modelling. Int J Dis Risk Reduct 81:103290–103290. https://doi.org/10.1016/j.ijdrr.2022.103290
Ómarsdóttir IL, Pétursdóttir G, Bird DK, Gísladóttir G (2022) Community resilience through the lens of social

support: recovering from the 2010 Eyjafjallajökull eruption in Iceland. Int J Dis Risk Reduct 81:103272–103272. https://doi.org/10.1016/j.ijdrr.2022.103272

Paschalia YPM, Doondori AK, Sekunda MS, Budiana I (2022) Revitalization of disaster mitigation management and community empowerment through community-based disaster risk reduction (CBDRR). J Penelitian Pendidikan IPA 8(5):2563–2574. https://doi.org/10.29303/jppipa.v8i5.2368

Prianto AL, Abdillah A (2023) Resilient cities, vulnerable communities: disaster governance in the coastal cities in Indonesia. In: International handbook of disaster research. Springer, Singapore, pp 1–12. https://doi.org/10.1007/978-981-16-8800-3_193-1

Quinn T, Adger WN, Butler C, Walker-Springett K (2021) Community resilience and well-being: an exploration of relationality and belonging after disasters. Ann Am Assoc Geogr 111(2):577–590. https://doi.org/10.1080/24694452.2020.1782167

Regmi NR, Giardino JR, Vitek JD (2010) Assessing susceptibility to landslides: using models to understand observed changes in slopes. Geomorphology 122(1–2):25–38. https://doi.org/10.1016/j.geomorph.2010.05.009

Reinhardt GY, Ross AD (2019) Expanding social science through disaster studies. Soc Sci Q 100(7):2523–2529. https://doi.org/10.1111/ssqu.12668

Robertson T, Docherty P, Millar F, Ruck A, Engstrom S (2021) Theory and practice of building community resilience to extreme events. Int J Dis Risk Reduct 59:102253–102253. https://doi.org/10.1016/j.ijdrr.2021.102253

Roy T, Nasreen M, Hasan MK, Sony MMAAM (2023) Community priorities in disaster risk reduction interventions. In: Coastal disaster risk management in Bangladesh. Routledge, Berlin, pp 335–355. https://doi.org/10.4324/9781003253495-21

Rözer V, Surminski S, Laurien F, McQuistan C, Mechler R (2023) Multiple resilience dividends at the community level: a comparative study of disaster risk reduction interventions in different countries. Clim Risk Manag 40:100518–100518. https://doi.org/10.1016/j.crm.2023.100518

Saghin D, Lupchian M-M, Lucheș D (2022) Social cohesion and community resilience during the COVID-19 pandemic in Northern Romania. Int J Environ Res Public Health 19(8):4587–4587. https://doi.org/10.3390/ijerph19084587

Shaw R, James H, Sharma V, Lukasiewicz A (2022) Disaster risk reduction and resilience: practices and challenges in Asia Pacific. In: Disaster risk reduction in Asia Pacific: governance, education and capacity. Springer, London, pp 1–15

Siders AR (2019) Adaptive capacity to climate change: a synthesis of concepts, methods, and findings in a fragmented field. Wires Clim Change. https://doi.org/10.1002/wcc.573

Sina D, Chang-Richards AY, Wilkinson S, Potangaroa R (2019) What does the future hold for relocated communities post-disaster? Factors affecting livelihood resilience. Int J Dis Risk Reduct 34:173–183. https://doi.org/10.1016/j.ijdrr.2018.11.015

Sousa C, Moss T (2022) Community resilience. In: Encyclopedia of social work. NASW Press and Oxford University Press, Oxford. https://doi.org/10.1093/acrefore/9780199975839.013.1441

Steffen W, Rockström J, Richardson K, Lenton TM, Folke C, Liverman D, Summerhayes CP, Barnosky AD, Cornell SE, Crucifix M, Donges JF, Fetzer I, Lade SJ, Scheffer M, Winkelmann R, Schellnhuber HJ (2018) Trajectories of the earth system in the Anthropocene. Proc Natl Acad Sci 115(33):8252–8259. https://doi.org/10.1073/pnas.1810141115

Sterlacchini S, Ballabio C, Blahut J, Masetti M, Sorichetta A (2011) Spatial agreement of predicted patterns in landslide susceptibility maps. Geomorphology 125(1):51–61. https://doi.org/10.1016/j.geomorph.2010.09.004

Suleimany M, Mokhtarzadeh S, Sharifi A (2022) Community resilience to pandemics: an assessment framework developed based on the review of COVID-19 literature. Int J Dis Risk Reduct 80:103248–103248. https://doi.org/10.1016/j.ijdrr.2022.103248

Supratiwi Y, Kushandajani D (2022) Community participation in disaster management in Gondoriyo village administration, Ngaliyan sub-district, Semarang city. IOP Conf Ser Earth Environ Sci 1041(1):012032–012032. https://doi.org/10.1088/1755-1315/1041/1/012032

Suryani YE, Darupratomo D, Ratnanik R, Anghraini KR, Ambarsari DW, Nisfa S (2022) Community participation in disaster risk reduction "Peseduluran Village". Soc Hum Educ Stud (SHES) Conf Ser 5(4):295–295. https://doi.org/10.20961/shes.v5i4.69101

Sutton SA, Paton D, Buergelt P, Sagala S, Meilianda E (2022) 'Village' as verb: Sustaining a transformation in disaster risk reduction from the bottom up. Environ Sci Policy 137:40–52. https://doi.org/10.1016/j.envsci.2022.08.009

Svyrydenko D, Nesterova M, Dielini M (2022) Changes in social cohesion in the university community in the context of crisis phenomena in society. Baltic J Econ Stud 8(3):185–196. https://doi.org/10.30525/2256-0742/2022-8-3-185-196

Szreter S (2004) Health by association? Social capital, social theory, and the political economy of public health. Int J Epidemiol 33(4):650–667. https://doi.org/10.1093/ije/dyh013

Tierney KJ (1999) Toward a critical sociology of risk Socio. Forum 14(2):215–242

Tong X, Zhang H (2020) A comprehensive analytical framework for risks, disasters and crises, pp 3–40. https://doi.org/10.1007/978-981-13-9140-8_1

Trut D, Kovačević J (2022) Climate change, disaster risk reduction and resilience. Environ Eng 9(1–2):35–42. https://doi.org/10.37023/ee.9.1-2.5

UNDRR (2019) United nations office for disaster risk reduction

UNISDR (2007) Building disaster resilient communities: good practices and lessons learned

UNITED NATIONS (2015) Sendai framework for disaster risk reduction 2015–2030

Van Niekerk D, Nemakonde LD, Kruger L, Forbes-Genade K (2018) Community-based disaster risk management. Handb Dis Res 2018:411–429

Virapongse A, Gupta R, Robbins ZJ, Blythe J, Duerr RE, Gregg C (2022) How can earth scientists contribute to community resilience? Challenges and recommendations. Front Clim. https://doi.org/10.3389/fclim.2022.761499

Willar D, Supit S, Kabo D, Senduk N (2022) Community resilience model: a case study of an earthquake disaster. Proc Int Struct Eng Constr. https://doi.org/10.14455/ISEC.2022.9(1).RAD-10

Willison ZD, Loreine BDLC, Luneta MD (2022) Community-based disaster risk reduction and climate change adaptation. In: An integrated governance approach towards a water-energy-food nexus and climate change. Edward Elgar Publishing, London, pp 330–352. https://doi.org/10.4337/9781800371613.00036

Wisner B, Blaikie P, Cannon T, Davis I (2004) At risk: natural hazards, people's vulnerability and disasters, 2nd edn. Routledge, Berlin

Wu J, Yang S, Wang W, Jaeger C (2023) How effective are community-based disaster reduction strategies? Evidence from the largest-scale program so far. Risk Anal 43(8):1667–1681. https://doi.org/10.1111/risa.14043

Xu W, Xiang L, Proverbs D, Xiong S (2020) The influence of COVID-19 on community disaster resilience. Int J Environ Res Public Health 18(1):88–88. https://doi.org/10.3390/ijerph18010088

Xu J, Zeng Z, Hong Y, Xi Z, Zhu X, Peng Z (2022) Grassroots mirroring under COVID-19: Does community resilience affect residents' responses? The case of Shenzhen, China. Sustainability 14(16):10159–10159. https://doi.org/10.3390/su141610159

Yang E, Kim J (2023) Sustainable tourism development in a host community: the mediating role of community resilience in response to disasters and crises. Int J Tour Res 25(5):527–541. https://doi.org/10.1002/jtr.2592

Yasukawa S, Delgado LMB (2023) Enhancing disaster preparedness and building resilience: UNESCO's multi-hazard, multi-disciplinary, and multi-stakeholder approach on disaster risk reduction, pp 93–100. https://doi.org/10.1007/978-981-19-5566-2_2

Zaman MO, Raihan MMH (2023) Community resilience to natural hazard induced-disasters: a systemic review of contemporary methods and theories. Nat Hazards Res 3(3):583–594. https://doi.org/10.1016/j.nhres.2023.05.003

Zander KK, Sibarani R, Abunyewah M, Erdiaw-Kwasie MO, Moss SA, Lassa J, Garnett ST (2023) Community resilience across Australia towards natural hazards: an application of the conjoint community resiliency assessment measurement. Disasters. https://doi.org/10.1111/disa.12590

Contextualizing Disaster Risk Reduction Through Digital Inclusion of the Vulnerable Groups at Yogyakarta, Indonesia

9

D. R. Hizbaron, B. H. Wibisono, S. Egaravanda, H. E. T. W. Nugroho, and A. S. Wahyu

Abstract

Contextualizing disaster risk reduction (DRR) is challenging not only for the government but also for civil society. From a social science perspective, disasters are understood not merely as events causing negative impacts, but as processes that create vulnerability within communities. This research posits that digital inclusion can reduce vulnerability by improving access to information and communication. Accordingly, it aims to contextualize DRR practices through the lens of digital inclusion among vulnerable groups in Yogyakarta, Indonesia. To achieve this, the study employed a Geographic Information System (GIS) to overlay statistical data on vulnerable groups—including children, the elderly, women and persons with disabilities—with spatial data on the availability of communication infrastructure. Using a generic spatial approach, the research examined the distribution of digital infrastructure in selected districts across Yogyakarta, as well as the characteristics of these vulnerable populations. The results revealed that Yogyakarta provides adequate digital infrastructure for vulnerable groups. However, further questions remain, such as how digital infrastructure is utilized before, during and after disaster events. Moreover, digital inclusion in Yogyakarta appears strongly associated with enhancing economic and social capital. The Yogyakarta experience suggests that efforts to contextualize DRR through digital inclusion are multifaceted and yield varying outcomes in terms of how communities experience and respond to disaster risks.

Keywords

Disaster · Risk · Vulnerable · Inclusion · Indonesia

D. R. Hizbaron (✉)
Department of Environmental Geography, Faculty of Geography, Universitas Gadjah Mada, Yogyakarta, Indonesia
e-mail: dyah.hizbaron@ugm.ac.id

B. H. Wibisono
Department of Architecture and Planning, Faculty of Engineering, Universitas Gadjah Mada, Yogyakarta, Indonesia

S. Egaravanda · H. E. T. W. Nugroho
Communication and Informatics Agency, Yogyakarta Special Province, Yogyakarta, Indonesia

A. S. Wahyu
Kampung Cyber Community, Yogyakarta Special Province, Yogyakarta, Indonesia

D. Hidayati et al. (eds.), *Examining Disaster Risk Reduction in Indonesia*,
Sustainable Development Goals Series, https://doi.org/10.1007/978-981-96-4182-6_9

9.1 Introduction

Contextualized disaster risk reduction has been globally recognized since early 2005 in Indonesia, right after the Tsunami Aceh, marked by the birth of Hyogo Framework for Disaster Risk Reduction. The idea to integrate disaster risk reduction into development planning became a national and international mandate to be manifested (UNDRR 2019). In fact, plenty of disaster risk reduction praxis is embedded within the civic society and originally initiated at the grass roots level (Cadag and Gaillard 2012; Suarma et al. 2018). The word contextualize here refers to the process of identifying disaster risk reduction according to context, such as the scale of observation, hazard origin and type of vulnerability and risk within the area of observation. Here, the research focuses on Yogyakarta Special Province, as an area exposed to various natural hazards, such as volcanic eruption, earthquake, landslide, drought and flood. The multi-risk map of Yogyakarta Special Province indicated that the urbanized area of Yogyakarta City, Sleman District and some of the area along the coastline of Kulonprogo District were exposed to higher multi-risk. The spatial distribution of disaster risk focused more on the urbanized area where people resided. Hence, strong disaster risk reduction is critical for Yogyakarta, especially in urban areas. How the digital inclusion embraces the vulnerable groups at Yogyakarta?

Disaster risk map is available in different formats of actions and digital actions. The digitalization involves the use of the Internet of Things (IoT), and Information Technologies (IT), which have been massively supporting disaster risk reduction. The previous research revealed that the digital activities during the disaster emergency had been enormous. The semantic sentiment analysis indicated from the big data crawling from Twitter that the type of expression during disaster events originated from the surrounding

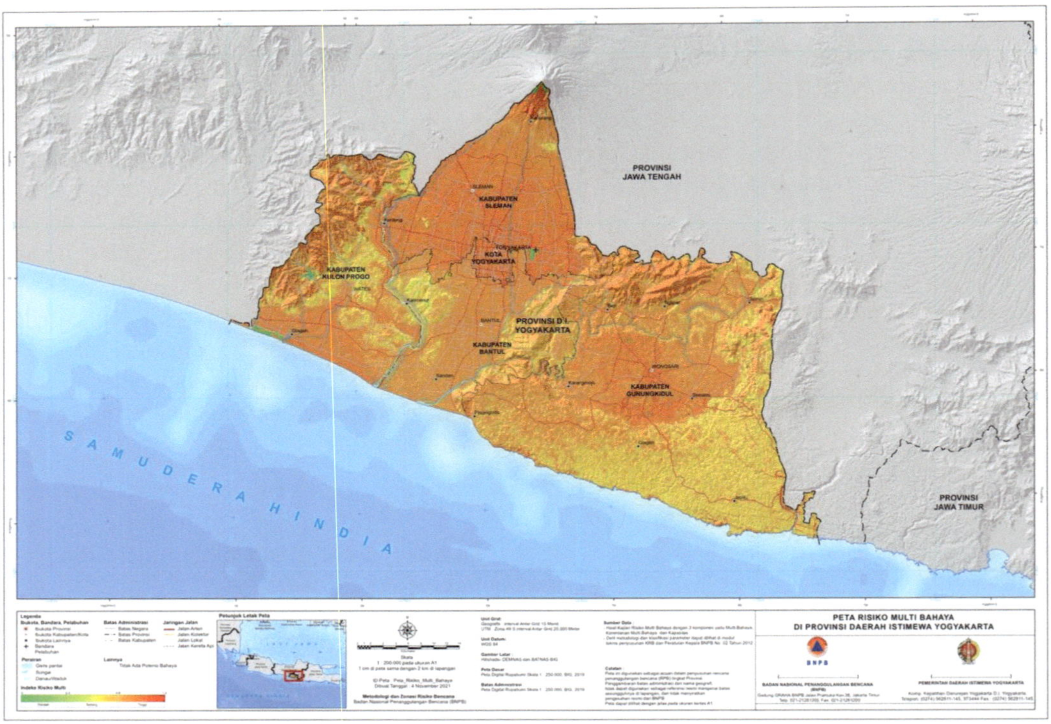

Fig. 9.1 Multi-hazard map of Yogyakarta Special Province (*Source* BNPB and Government of Yogyakarta Special Province, 2019)

areas, provided information about impacted areas and or simply showed sympathy (Salsabilla and Hizbaron 2021; Widyanarko and Hizbaron 2020). The digital expression at certain periods also enables scholars to monitor the scale of disaster occurrence during emergencies such as COVID-19 pandemics (Hizbaron et al., 2021). Previous research also argued that through the combination of disaster statistics, system information on disaster risk and spatial planning, three pivotal urban areas in Sulawesi have evidently shown rapid development planning at the highest disaster risk areas (Hizbaron et al. 2023). Here, the use of IoT, IT and digital information has been widely implemented to learn pre-disaster, emergency and post-disaster events.

Aside from the emergency situation, digitalization contributes so much to the pre-disaster events and post-disaster events in Indonesia, such as disseminating disaster-related statistics as indicated by websites: https://dibi.bnpb.go.id and https://gis.bnpb.go.id. The geospatial data information about the likelihood of hazard, vulnerability, and risk information is also available on https://inarisk.bnpb.go.id. Each province and district across Indonesia tries to provide accessible disaster management information using the websites of the Local Disaster Management Agency. Similar to the idea, The Ministry of Environment and Forestry Republic Indonesia established a web-based system for information on climate change vulnerability or Sistem Informasi Data Indeks Kerentanan (SIDIK), which can be accessed through http://sidik.menlhk.go.id. In larger contexts such as Asia Pacific, the United Nations Economics and Social Commission for Asia Pacific (UNESCAP) also promotes the use of web-GIS portal to describe the hazard hotspot and potential impacts in the member countries as presented from https://rrp.unescap.org. The digitalization is not limited to the use of websites and geoportal per se; it also engages with the use of digital platforms such as Android Apps-based platforms or IoS Apps-based platforms. Herewith, the Indonesian government also provided InaRisk Apps using an Android-based platform to increase community awareness about disaster risk. As the culture to access digital information has been globally known, it is mandatory for the government to provide information in advance, including prediction scenarios of disaster risk. The government also provides digital infrastructure to the region for public use. How is the distribution of the digital infrastructure in Yogyakarta urban areas?

As the research tries to identify the use of digital infrastructure and its spatial distribution, at the other continuum, this research also examines the process of digital inclusion from grass roots level. The idea is to collect shared understanding about digital inclusion for the vulnerable groups at Yogyakarta being taken into account.

9.2 Research Context

Resilience is a buzzword when we discuss urban slums, urban poor, urban disaster and or urban climate change. In the emerging economics nowadays, the global challenge has impacted in many ways, especially vulnerable groups such as the poor, or the least access to urban facilities. During these days, digitalization emerged as an entity of daily business, starting from administrative matters to information dissemination, which is crucial for before, during and right after the disaster event. The necessity to access nearly real-time and real-time data has been a massive demand all over the globe. The remaining question here is, how do we equip the digital information to minimize urban vulnerability and to attain resilient community?

Profoundly, the vulnerable groups are understood as a collective number of specific groups of the population (Cutter 2017; Cutter et al. 2003; Hizbaron et al. 2021a). Here, the indicator for the vulnerability includes groups of elderly, children, women and poor households. Argued from geography science, using spatial science approach, these vulnerable groups mostly account for accumulated number or proxies in particular locations that are exposed to particular hazards or exposures (Bankoff et al. 2004; Guillard-Gonçalves and Zêzere 2018). In social science, using a participatory approach, these vulnerable groups often articulated to confront a very specific situation that caused them vulnerable (Arora-Jonsson 2011; Dougherty et al. 2016). Thus, the idea is to explore case by case each individual considered vulnerable and route back to the main or root cause that influences their current state. Here, both scientific domain offers similar perspective that vulnerability is a contextual framework to understand one's condition due to particular inability to confront the exposure or hazards. The differences lie within the method of doing research. The applied science heavily relies upon the spatial approach, so providing quick scanning for large-scale areas is very useful. Meanwhile, social science is generally approached using case-by-case or participatory approach. Therefore, it is very useful to identify a very specific case study. The research presented here is a combination of both approaches.

The understanding of digital inclusion here meaning very broadly. The digital inclusion refers to the inclusion of knowing, understanding and comprehending the digital data using adequate access to digital infrastructure and promoting responsive behavior to disaster-related events. Nowadays, we often work with big data due to massive use of Internet of Things and improved innovation technology. The big data generally originated from satellites, global positioning system (GPS), smartphones, sensors and many other devices (Ferreira and Vale 2022; Kitchin 2014). The volume of big data is enormous, we can easily subtract the data from Internet in different formats of data through the internet. An example of big data is a series of tweet data (semantic), Instagram photos (visual), Facebook information (numeric and semantic), web-GIS portal (nodes, location, GPS, temporal) and many more. The idea to work with digital data from the internet provides two different sets of data, such as nearly real-time and real-time data. Massive nearly real-time and real-time data are important for disaster research since the emergency phase relies heavily upon quick response. The quick response to disaster also requires a quick decision-making process, which is supported by adequate infrastructures. Data transformation has become a turning point in science, including Geographic Information Systems. Goodchild et al. 2007 argued that there are new general representations of information derived from such huge data availability. At some points, the available information helps people make decisions, in times of emergency or during business as usual (Ferreira and Vale 2022; Goodchild et al. 2007; Kitchin 2014).

The idea of the research is to observe digital inclusion in disaster risk reduction. The observation is conducted in two units of analysis using two approaches. First, the research observes the distribution of vulnerable groups across Yogyakarta Greater Areas. To evaluate the availability of digital infrastructure across the Yogyakarta Greater Area, the research conducted spatial statistical analysis. The spatial statistics aim to compare two indicators: vulnerable groups and digital infrastructures. Second, the research thoroughly looks at smaller units of analysis, which is Kampung Cyber, Yogyakarta. The Kampung Cyber portrays successful lessons learned in providing digital inclusion to the community. The qualitative approach is implemented within this step to pursue the complete narratives upon the indicators of digitalization, accumulated capital, stakeholder involvement and supporting elements that accelerate the digital inclusion process (Fig. 9.2).

The data considered was collected from the Communication and Informatics Agency of Yogyakarta Province. The remaining data from the population statistics is referred to the Statistical Bureau of Yogyakarta Province. The collected data was hence analyzed using spatial approach to understand the overlapping pattern between the infrastructure distribution and population characteristics. Once the larger scale is portrayed through the spatial analysis, this research explores qualitatively through in-depth interview for the Kampung Cyber perspective. The idea is to put the experience of the local community in providing better collective aspects in digitalization and the use of Internet of Things.

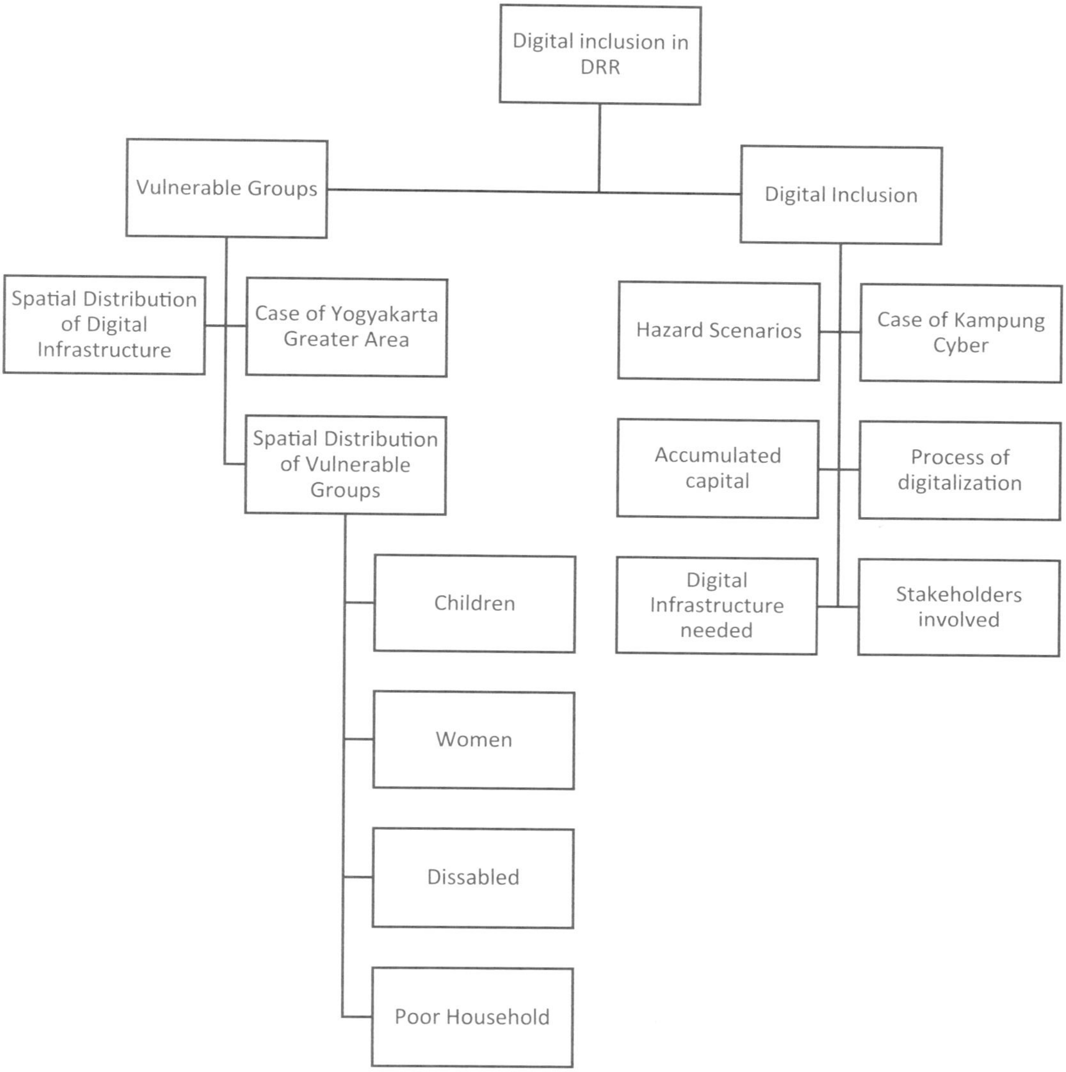

Fig. 9.2 Research framework

9.3 Research Result and Discussion

A discussion to vulnerability often relates with the existing threat, either hazard or susceptibility from natural occurrence. For example, a vulnerable group to earthquake refers to their potential inability to confront potential earthquake occurrence (Bankoff et al. 2004; Hizbaron et al. 2012, 2018, 2023). Here, the research understands vulnerability in Yogyakarta as the inability to be exposed to various hazards, such as earthquakes, volcanic eruptions, floods, typhoons, landslides and tsunamis. The Greater Yogyakarta, as indicated in Fig. 9.3, is currently exposed to earthquakes, volcanic eruptions and typhoons. Arguably, the digital infrastructures are ideally available to assist during pre-disaster events, per-disaster events or emergency situations and post-disaster events.

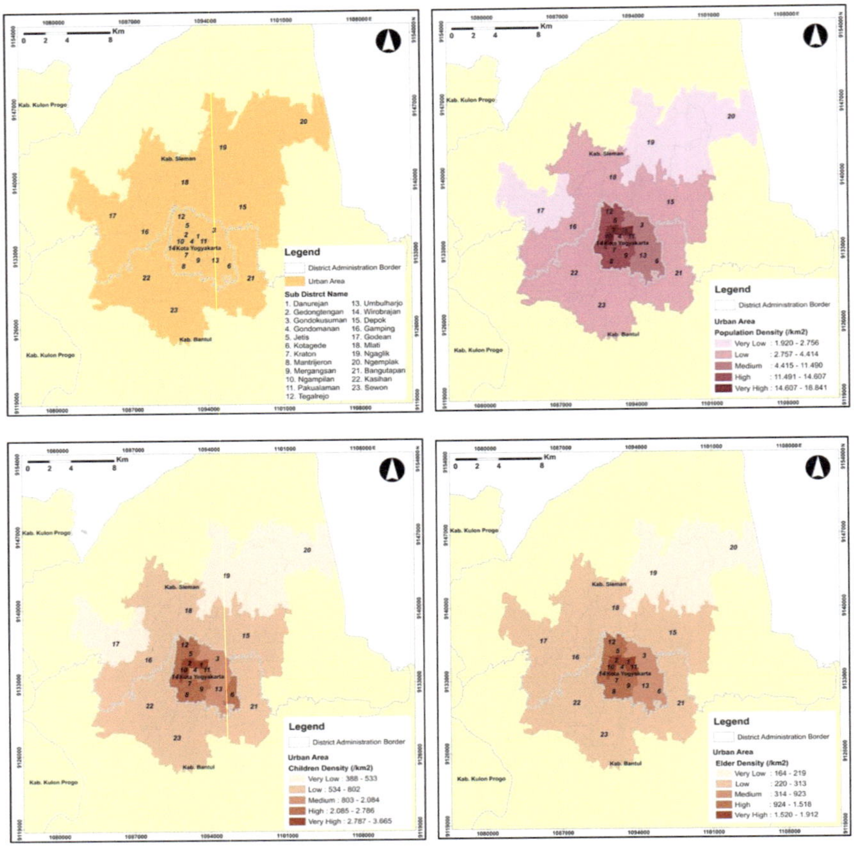

Fig. 9.3 The Greater Yogyakarta population, BPS, 2023

The Governor of Yogyakarta sets three important visions within the regional development plan, such as the development of the southern region, the village reformation and the technology information implementation. Statistically, the local government admitted to not fully covering the digital infrastructure facilities due to the availability of blank spots. However, currently the local government has installed fiber optic for 500 km alongside the regions. Meanwhile, the actual funding to establish the digital infrastructure for Yogyakarta Special Province is approaching 9 million IDR per annum, while for Yogyakarta Municipality is 5 million IDR per annum. The government support for developing the Yogyakarta area, especially the southern part has been clearly stated. Hence, the research tries to observe how the distribution of the digital infrastructure matched with the important visions to the regional development of the region.

The research is conducted at Greater Yogyakarta Area (GYA) as the urbanized part of Yogyakarta Province. Figure 9.3 indicates that the Yogyakarta Greater Area comprises The City of Yogyakarta, partially Sleman Districts, and Bantul Districts. There are 23 sub-districts across Yogyakarta Greater Area. The population density, children and elderly are accumulated at north-western part of Yogyakarta City. Meanwhile, the distribution of poor household concentrated outside Yogyakarta City and accumulated to the southern part of Bantul District. There are 23 sub-districts considered as urban areas, whereas 14 of the sub-districts are concentrated at the Yogyakarta City. The most densely populated area accumulated at the sub-district at Yogyakarta City which accounted up to 14,000–15,000 lives

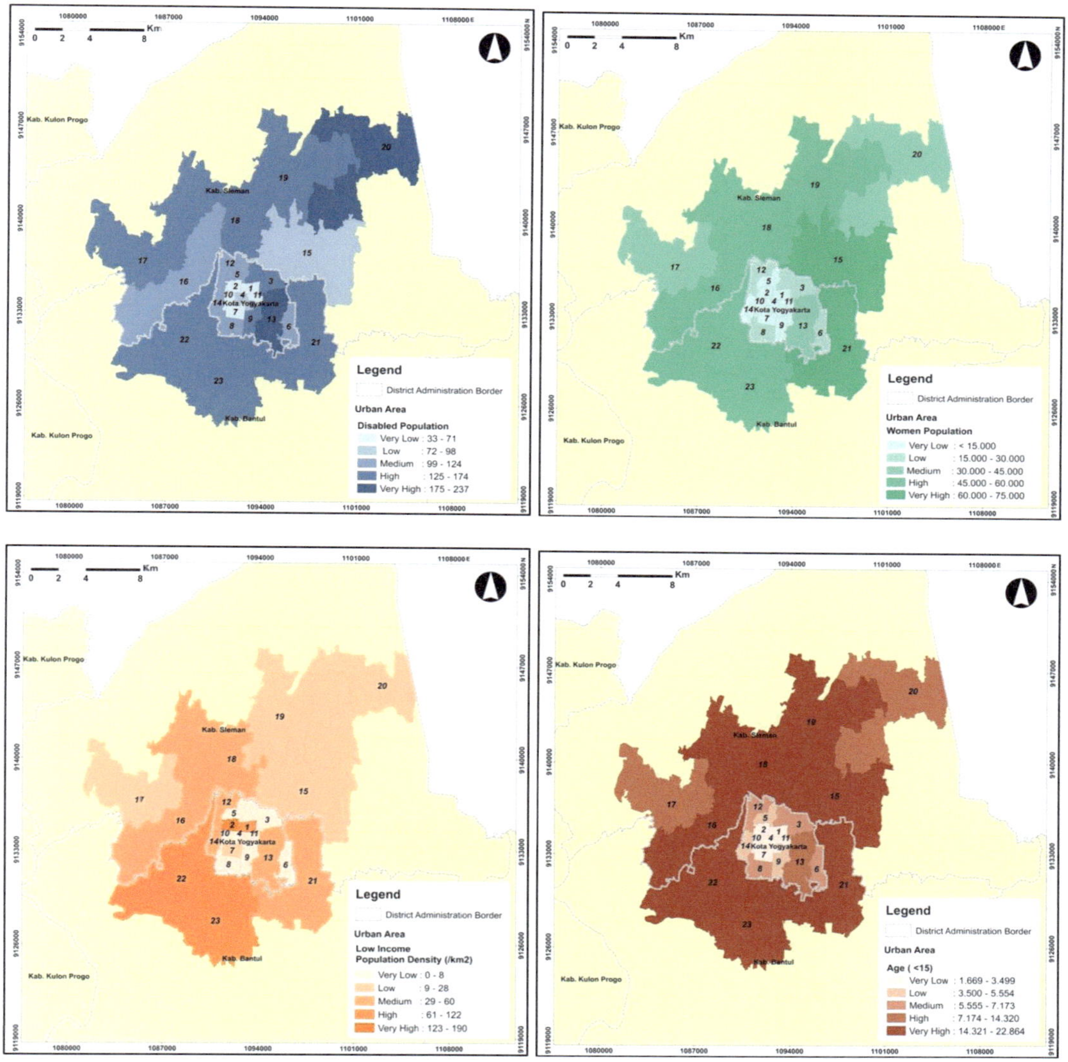

Fig. 9.4 The Greater Yogyakarta vulnerability characteristics, BPS, 2023

per km^2. The most potentially vulnerable groups within the GYA are predominantly children and elderly are concentrated at the city center of Yogyakarta.

Slightly different to the previous likelihood, the outskirts of Yogyakarta City is occupied by vulnerable groups especially disabled community, women, non-productive ages below 15 and poor households (Fig. 9.4). Reflecting to the hazard distribution, we learned that most of the children and elderly are mostly exposed to the typhoon and seismic hazards. Meanwhile, the women, disabled community, non-productive ages below 15 and poor households are commonly exposed to volcanic eruption (Sleman District), and seismic potentials (Bantul and Yogyakarta). A specific concern to the disabled community, during emergency situation normally confront with numerous bulk issues. The bulk issues are neglected during evacuation, displacement and return home (Twigg et al. 2018). This is essential to ensure the disable group to have adequate access to informa-

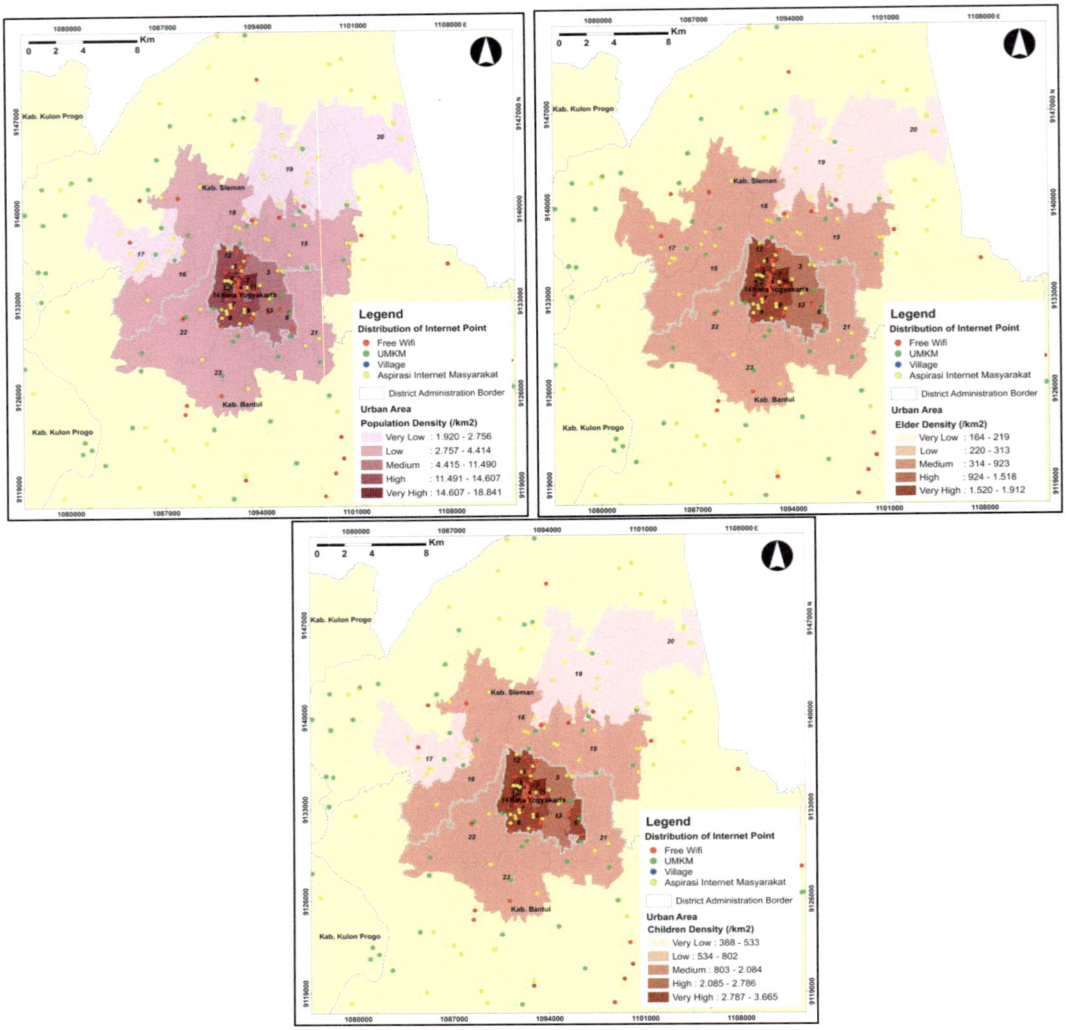

Fig. 9.5 The Greater Yogyakarta digital infrastructure characteristics, 2023

tion during emergency situation. An appropriate handling to the disabled community during emergency situation has always been a struggle. Hence, the optimum digital inclusion to the disable group is a manifestation to get away from the bulk issue by making the best use of information for further actions during emergency situation.

This research takes a closer look the digitalization from two perspectives; first is the digital infrastructure and digital data. The digital infrastructure provided by the Provincial Government of Yogyakarta is distributed across Yogyakarta Special Province. As argued earlier, the idea of the provincial government is to support access to digital inclusion to the very small-scale unit, such as village level. However, Fig. 9.5 below indicates some haphazard to the main idea. As also explained above, the haphazard is also due to the location where blank spots exist, as well as financial hurdles. Figure 9.5 also indicates that there are four types of internet infrastructure points provided through free wi-fi from local government, private-owned unit, village initiatives and collective initiatives.

The urbanized Yogyakarta City gets very dense support upon free wi-fi access from the local

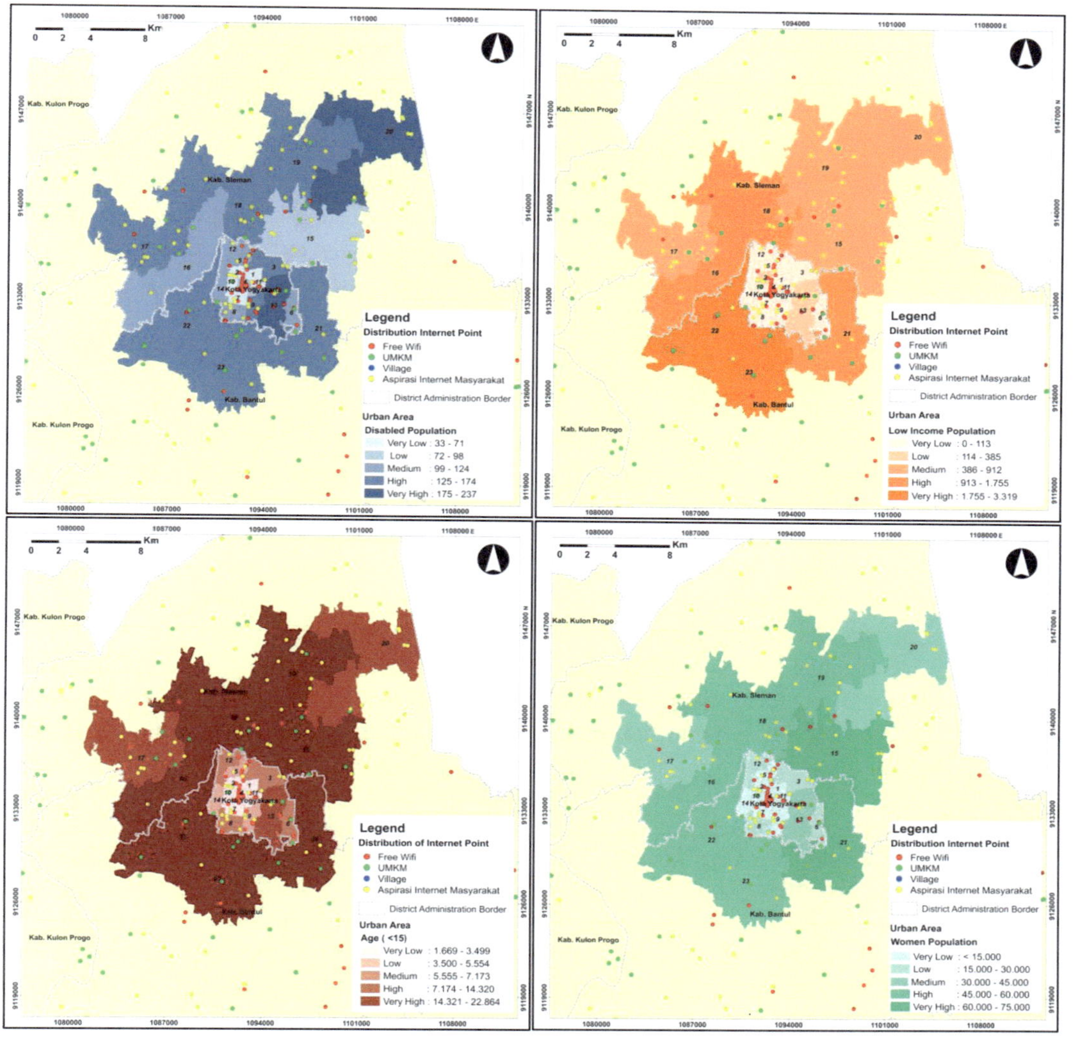

Fig. 9.6 The Greater Yogyakarta vulnerable groups and the digital infrastructure distribution, 2023

government and connects with very intense collective initiatives. In the area where vulnerable groups such as children are accumulated, the city has been supported tremendously with wi-fi access from the local government and private owned. The Greater Yogyakarta Area has been fully supported with dense wi-fi connection. Here, we can argue that the urbanized area has been very much accessible to any digital inclusion. Although, the accumulation of vulnerable groups is enormous at the city center, the availability to access real-time data through any kind of digital infrastructure is inevitably profitable for the vulnerable groups. The access to real-time or nearly real-time information before, during and after disaster occurrence is presumably easier compared to those area provided with less digital infrastructures.

The following Fig. 9.6 indicates that the disable population mostly supported with various digital infrastructures at the vicinity of Yogyakarta City. However, those disable population resided at the Sleman District have less access to the availability of digital infrastructure. From here, we can pick up the logic that the inclusion of disable groups in the disaster risk mitigation through increasing their access to information is important thing to be done (Twigg et al. 2018).

The distribution of low-income group also spatially heterogeneous in terms of getting support from the digital infrastructure. Most of the facilities are supporting the vulnerable groups at the southern part of Yogyakarta, which is Bantul District. Meanwhile, some of the poor income groups at the area of Sleman District are supported with less infrastructure. The productive age here is not considerably vulnerable group compared to the elderly or children. However, considering their location where hazards likelihood existed, their existence is also considerably vulnerable. Most of the productive age have very good access equally at any areas. Meanwhile, the women distribution here get very nice access to the digital infrastructure. To argue, as the local regional development is centered to confront the southern area as the main gate, the distribution or equipment to the digital infrastructure to the southern part of Yogyakarta should be increased. This will be a very strong commitment to support the development and also minimize the vulnerable groups especially the low-income group. The resilient community is essential to be created from the existence of adequate digital infrastructure.

The distribution of vulnerable groups such as elderly for example, has been seen to get modest support from the wi-fi access. The densely populated vulnerable group for elderly focused at the western part of Yogyakarta City, where the support of free wi-fi is tremendous. The range of access is decreasing to the outskirts of the city center. The spatial pattern for vulnerable groups such as elderly and children is similar; thus, vulnerable groups at the city center are able to get better access to digital inclusion compared to those resided at the outskirts of Yogyakarta city. Here, we argue that even in the Greater Yogyakarta Area, there are slight distinctions to the existence of digital inclusion in terms of the infrastructure.

In order to support the argument that the Yogyakarta is potentially able to provide private-owned wi-fi, the research also presents the qualitative observation to cyber-village profile. The "Kampung Cyber" or cyber village is located at Kraton sub-district, which is situated at the very heart of the Yogyakarta City. The Kampung Cyber was established in 2008 through very collective efforts.

One of its key person is the local resident, who later acted as the Neighborhood Headmaster for the area. The main idea was to provide internet network for all communities. The Kampung Cyber is likely conducted in a very common small alley kind of kampung, as it reflected a very common urbanized neighborhood in Yogyakarta, where people live in a very densely located residential areas as shown in the following Fig. 9.7.

Back in 2008, the internet connection was limited, therefore, the kampung could not easily

Fig. 9.7 The Greater Yogyakarta vulnerable groups and the digital infrastructure distribution, 2023

access information via internet. Thus there were three important challenges at the first establishment of Kampung Cyber, such as inadequate PCs (Personal Computers), the supporting infrastructure to internet was not economically feasible, and very limited know-how-to on technical set up to the internet. By education the community had less understanding to the use of technology, therefore, the local community had no idea to the importance of the internet whatsoever. Hence, the internet was provided for public in very limited number of PCs, and bandwidth due to limited financial support. The local headmaster introduced the importance of the internet, which was challenging at that time. As the main livelihood source was from informal sectors, the challenges were tackled with training for local community to use internet and to increase day-to-day discussion via community events, guard post discussion and many other media to inform the internet use. Afterward, the local community creates networking cables to connect each houses, where all the costs were borne by the community. For 7 years, the bandwidth and all infrastructure were increasing a lot. The next step of the use of the internet is to use the internet for small medium enterprise marketing support, population administrative matters and many others. The kampung cyber also creates an application to support the implementation of internet. Nowadays, the Kampung Cyber has all full access to internet, with several public wi-fi provided for public use. The community is also able to set up their small-scale medium enterprise campaign through the web and creates livelihood out of it. The local community is also very famous to serve all tourist who are interested to learn art and culture in Yogyakarta through online-training. During disaster occurrence, the local community is also able to access quick information from the internet and enables them to conduct mitigation strategy. The case of Kampung Cyber is an example on how the collective idea is being implemented and creates stronger social capital which supports the urban resilience.

9.4 Conclusion

During the pandemic, the world was exposed to higher vulnerable groups due to declining economic (9.2% of world population lived under poverty lines and 53% dependency rate hit hardest by COVID-19 pandemic), and increasing global exposure due to disaster.[1] The global economic problem, climate change adaptation and disaster risk mitigation failure are the top-listed future challenges that need to be confronted. In terms of digital infrastructure there are several international commitments guidelines must take place, i.e.:

1. Disaster data digitalized and integrated using cloud-based technologies to support disaster risk reduction and recovery in project countries;
2. Targeted digital solutions for increased preparedness and response of vulnerable groups developed and applied;
3. Strengthened capacities of national and subnational governments through partnerships for disaster risk-informed development planning.

Reflecting from the global phenome and commitment, we learned from Yogyakarta case, that the existence of vulnerable groups should not limited our effort to identify the existence of our capacity to confront disaster risk. Here, we learned that the urban vulnerable groups have been equipped with spatially heterogeneous supports to information with the establishment of the private-owned, collective and public digital infrastructures. The spatial approach enables us to identify at the same time the distribution of vulnerability and capacity to tackle disaster risk at once.

The Yogyakarta Province at large is also supported by digital infrastructure which is mainly supported by private-owned wi-fi facilities especially to the north, west and southern parts of the area. The government support to free wi-fi access

[1] https://www.preventionweb.net/news/connect-innovate-accelerate-digital-solutions-disaster-risk-reduction-crisis-response.

mainly accumulated at Yogyakarta City. Meanwhile, the area of eastern part of Yogyakarta is mainly supported by collective initiatives wi-fi access. From here we learn that the digital inclusion from infrastructure perspective is twofold. The improvement to digital information is equally important at the northern part of Sleman District especially to support the disable group. Additionally, room for improvement to the availability of digital infrastructure is also possible to be prioritized at the southern part of Bantul Districts where the poor income groups are accumulated.

First, the infrastructure for digitalization is borne not only by the local government, but also vested from private and collective initiatives. Derived from the fact finding, this research learned that the Yogyakarta case has provided profound support or initiated a very strong initiatives to social capital upon the digital infrastructures availability. The understanding to initiate the private-owned and collective wi-fi means as a fundamental capital to achieve better support to recent information through the internet connection. The finding similar to the previous arguments stating that the vulnerable groups also potentially increase their capacity to minimize further impact to disaster occurrence through different mechanism (Fauziyanti and Hizbaron 2020; Hizbaron et al. 2021b; Marfai and Hizbaron 2011).

Here, the research learned that one of the mechanisms to minimize disaster risk, the vulnerable groups in Yogyakarta are able to increase their capacity to digital infrastructure for better and updated information. Second, the distribution of digital infrastructures is mainly accumulated at the city center, while the rest of the area is supported from disperse source of support. Arguably, we also learned through the finding that the capacity nested in the community also spatially varied. The idea and initiatives to have private and collective means of infrastructure also sporadically exist. Therefore, a spatial approach is well implemented in the research to identify the spatial distribution of local capacity to minimize their vulnerability.

The research has elaborated using spatial approach to the digital infrastructure inclusion for Yogyakarta. However, the research has not yet elaborated on the type of digital data which are prominent to minimize disaster vulnerability for Yogyakarta. As in future, the world will be easier to access the digital data, thus we need to also raise a question, what type of digital data that increase our urban resilience? This will need further research.

References

Arora-Jonsson S (2011) Virtue and vulnerability: discourses on women, gender and climate change. Glob Environ Chang 21(2):744–751. https://doi.org/10.1016/j.gloenvcha.2011.01.005

Bankoff G, Frerks G, Hillhorst D (2004) Mapping vulnerability: disaster, development and people. Earthscan, London

Cadag JRD, Gaillard JC (2012) Integrating knowledge and actions in disaster risk reduction: the contribution of participatory mapping. Area 44(1):100–109. https://doi.org/10.1111/j.1475-4762.2011.01065.x

Cutter SL (2017) The forgotten casualties redux: women, children, and disaster risk. Glob Environ Chang 42:117–121. https://doi.org/10.1016/j.gloenvcha.2016.12.010

Cutter SL, Boruff BJ, Shirley WL (2003) Social vulnerability to environmental hazards. Soc Sci Quart 84(2):242–261

Dougherty S, Taylor J, Hidayani R, Fatimah D (2016) Climate change vulnerability assessments in Indonesia: Where are the women's perspectives? In: Asian cities climate resilience

Fauziyanti NU, Hizbaron DR (2020) Sustainable livelihood strategies: How urban community resilient towards disaster? Indones J Geogr 35(2):246–259. https://doi.org/10.22146/IJG.52412

Ferreira D, Vale M (2022) Geography in the big data age: an overview of the historical resonance of current debates. Geogr Rev 112(2):250–266. https://doi.org/10.1080/00167428.2020.1832424

Goodchild MF, Yuan M, Cova TJ (2007) Towards a general theory of geographic representation in GIS. Int J Geogr Inf Sci 21(3):239–260. https://doi.org/10.1080/13658810600965271

Guillard-Gonçalves C, Zêzere JL (2018) Combining social vulnerability and physical vulnerability to analyse landslide risk at the municipal scale. Geosciences (Switzerland). https://doi.org/10.3390/geosciences8080294

Hizbaron DR, Baiquni M, Sartohadi J, Rijanta R (2012) Urban vulnerability in Bantul district, Indonesia-towards safer and sustainable development. Sustainability 4(9):2022–2037. https://doi.org/10.3390/su4092022

Hizbaron DR, Hadmoko DS, Mei ETW, Murti SH, Laksani MRT, Tiyansyah AF, Siswanti E, Tampubolon IE (2018) Towards measurable resilience: mapping the vulnerability of at-risk community at Kelud Volcano,

Indonesia. Appl Geogr 97:217–227. https://doi.org/10.1016/j.apgeog.2018.06.012

Hizbaron DR, Ismayani N, Ernawan FN, Puspitasari NNA, Yulianda Y, Jati R (2021a) Comparative urban capacity analysis for flood-prone areas: method and a case study of Sidoarjo, Indonesia. Environ Develop Sustain. https://doi.org/10.1007/s10668-021-01249-8

Hizbaron DR, Ismayani N, Ernawan FN, Puspitasari NNA, Yulianda Y, Jati R (2021b) Comparative urban capacity analysis for flood-prone areas: method and a case study of Sidoarjo, Indonesia. Environ Dev Sustain 23(9):14027–14045. https://doi.org/10.1007/s10668-021-01249-8

Hizbaron DR, Dimyati M, Hadi MP, Khakim N, Suarma U, Pujiastuti I (2023) Co-creation of urban hazard, vulnerability and risk information system towards sustainable regional development trajectories in Sulawesi, Indonesia. Nat Hazards. https://doi.org/10.1007/s11069-023-06234-0

Hizbaron DR, Ruslanjari D, Mardiatno D (2021) Amidst covid-19 pandemic: an adaptive disaster governance in Yogyakarta, Indonesia. Social Sci 10(3)

Kitchin R (2014) Big data, new epistemologies and paradigm shifts. Big Data Soc. https://doi.org/10.1177/2053951714528481

Marfai MA, Hizbaron DR (2011) Community's adaptive capacity due to coastal flooding in Semarang Coastal City, Indonesia. Analele Universitat Din Oradea Serie Geografie, XXI(2), 209–211. http://istgeorelint.uoradea.ro/Reviste/Anale/anale.htm

Salsabilla FP, Hizbaron DR (2021) Understanding community collective behaviour through social media responses: case of Sunda Strait Tsunami, 2018, Indonesia. E3S Web Conf 325:01021. https://doi.org/10.1051/e3sconf/202132501021

Suarma U, Hizbaron DR, Nurjani E (2018) Participatory implementation within climate change related policies in urbanized area of Indonesia. Print Indones J Geogr 50(2):121–132. https://doi.org/10.22146/ijg.36262

Twigg J, Kett M, Lovell E (2018) Briefing note key messages disability inclusion and disaster risk reduction overcoming barriers to progress. https://interagencystandingcommittee.org/iasc-task-team-inclusion-persons-disabilities-humanitarian-action

UNDRR (2019) United Nations office for disaster risk reduction

Widyanarko U, Hizbaron DR (2020) Urban response towards tropical cyclone using twitter in Indonesia. IOP Conf Ser Earth Environ Sci. https://doi.org/10.1088/1755-1315/451/1/012098

Wisner C, Nivaran D (n.d.) At risk: natural hazards, people's vulnerability and disasters, 2nd edn

10 Adaptive Social Protection (ASP) for Post-disaster Recovery in Indonesia: Building Resilience and Inclusion Through Intersectional Framework

Yanu Endar Prasetyo, Syahmida S. Arsyad, and C. Yudilastiantoro

Abstract

Indonesia's diverse geography and high population density pose unique disaster management and recovery challenges. The country is home to over 17,000 islands, making the logistics of recovery efforts incredibly complex. Many of the population resides in vulnerable coastal areas, increasing their exposure to various natural hazard-induced disasters (NHID). Disasters, whether artificial or natural, disrupt people's lives, infrastructure, economies, and social systems and frequently have a significant and enduring impact on communities. Initiatives for reconstruction and rehabilitation are crucial during the crucial post-disaster recovery phase to regain stability and enable the afflicted communities to lead regular lives again. A post-disaster recovery strategy must include implementing adaptive social protection measures since they promote inclusivity and resilience-building. The recovery efforts must ensure inclusivity, focusing on vulnerable groups to mitigate the widening socio-economic gap. Adaptive social protection programs, skill development initiatives, and livelihood support are essential to a recovery strategy that addresses the needs of the most affected. Social sciences play a key role in post-disaster recovery by offering crucial insights and understanding of human behavior, societal dynamics, and the effects of disasters on individuals and communities. These insights guide the creation and implementation of effective recovery and reconstruction strategies, particularly in the Indonesian setting.

Keywords

Adaptive social protection · Post-disaster recovery · Resilience · Inclusion · Social sciences

10.1 Post-disaster Recovery in Indonesian Context

Indonesia, a nation prone to Natural hazard-induced disasters (NHID) due to its geographical location, faces significant challenges in post-disaster recovery (Cazabat et al. 2023). The archipelago is situated on the Pacific Ring of Fire, making it susceptible to earthquakes, tsunamis, volcanic eruptions, and floods (Fakhruddin and Elmada 2022). Indonesia's diverse geography and high population density pose unique challenges. The country is home to over 17,000 islands,

Y. E. Prasetyo (✉) · S. S. Arsyad · C. Yudilastiantoro
Research Center for Population, BRIN, Jakarta, Indonesia
e-mail: yanu005@brin.go.id

D. Hidayati et al. (eds.), *Examining Disaster Risk Reduction in Indonesia*,
Sustainable Development Goals Series, https://doi.org/10.1007/978-981-96-4182-6_10

making the logistics of recovery efforts especially complex (Gunawan et al. 2016). Moreover, a large portion of the population resides in vulnerable coastal areas, increasing their exposure to various forms of NHID (Almutairi et al. 2020).

There were some significant disasters in Indonesia. In 2004, the Sumatra earthquake and ensuing tsunamis resulted in the deaths of approximately 230,000–280,000 people across multiple countries (Syamsidik et al. 2023), making it one of the deadliest NHID in recorded history (Hagan et al. 2007). Tens of thousands of individuals were injured, many of whom suffered severe injuries requiring immediate medical attention (Rofi et al. 2006). The Bantul earthquake on May 27, 2006, resulted in the loss of 6234 lives, displacement of 46,148 individuals, and injuries to 26,299 people, both severe and mild (Elnashai 2007). It devastated thousands of buildings, including households and places of worship, totaling around 390,077 structures leveled to the ground. Specifically, in the Bantul area, 4143 people lost their lives (Nurwihastuti et al. 2014).

A mud volcano eruption in Sidoarjo, East Java, has been ongoing since 2006, displacing thousands of people and submerging homes and villages in mud (Agustawijaya 2013). The mud volcano eruption has caused significant land subsidence and surface displacement in the affected area, leading to changes in the landscape and potential damage to infrastructure (Plumlee et al. 2008). Mount Merapi, one of Indonesia's most active volcanoes, experienced significant eruptions in 2010 and 2018 (Yudistira et al. 2020), resulting in casualties, the displacement of communities, and damage to infrastructure (Mei et al. 2016).

In 2018, the country was hit by a devastating earthquake in Lombok and tsunami in Sulawesi, leading to significant loss of life and property. The series of earthquakes in Lombok resulted in the deaths of more than 550 people and injured thousands, causing significant loss of life and health impacts (Zhao et al. 2021). The earthquake and subsequent tsunami in Sulawesi resulted in the deaths of over 4000 people and caused injuries to thousands more (Supendi et al. 2020).

Following the disasters, various relief and rehabilitation efforts were undertaken. Numerous recovery and reconstruction programs were implemented to rebuild infrastructure and aid affected communities in recovering from the tragedy (Wagstaff et al. 2019). The earthquake also prompted increased awareness and preparedness for NHID, including education on evacuation procedures and the construction of earthquake-resistant buildings (Nefianto 2023). Indonesia already has various social protection programs aimed at addressing poverty and social vulnerability by enhancing and improving the population's capacity to protect themselves from disasters and income loss. In fact, the government of Indonesia realized the need for a more Adaptive Social Protection (ASP) system to support affected communities during and after disasters (Fitrinitia et al. 2022).

The Indonesian government has made efforts to implement structured and planned strategies and command posts (Ruswandi and Maarif 2019). The disaster management strategy begins at the end of the emergency response phase, involving damage and loss assessment or post-disaster assessment and calculation (JITUPASNA) (Jibiki et al. 2020). The result is an estimate of damage and losses calculated from five main sectors: housing, public infrastructure, socio-cultural, economic, and cross-sectoral. Subsequently, a Post Disaster Needs Assessment (PDNA) is conducted (Bollin and Khanna 2007), considering the needs for post-disaster recovery, both in terms of stimuli and assistance for housing rehabilitation and reconstruction, the need for assistance in the recovery of public facilities and infrastructure, governance, and socio-cultural aspects, as well as the need for assistance in economic recovery in the form of subsidies and incentives, indirectly aimed at medium and long-term recovery of losses (Behera and Das 2023).

Besides the comprehensive post-disaster handling by the government, an approach that can be emulated to address disasters comprehensively is the role of the community. They play a crucial role in pre-disaster, during-disaster, and post-disaster phases. Community activities in disaster-prone areas in the pre-disaster phase include participating in disaster prevention with the Local Government, engaging in mitigation efforts, attending disaster management training and socialization, and establishing Disaster-Resilient Villages.

Community activities during a disaster involve: providing information to the local Regional Disaster Management Agency (BPBD) or relevant institutions, self-evacuation, and participating in emergency response according to their expertise. Community activities post-disaster include participating in rehabilitation actions, reconstructing public infrastructure, especially roads and dwellings, and providing support for psychological recovery. Psychological counseling involves providing support and guidance to individuals and communities affected by disasters. This includes offering emotional support, assisting impacted individuals in accessing resources and services, and facilitating efforts for recovery and rehabilitation.

10.2 Understanding ASP from Intersectional Perspectives

Indonesian government adopted ASP approach to improve resilience and responsiveness to disasters (Gasior et al. 2023). The government integrated the following key elements in the system such as: (1) invested in advanced early warning systems, enabling timely evacuation and disaster preparedness; (2) the government integrated various data sources, including geographic information and household profiles, to identify vulnerable populations; (3) introduced cash transfer programs that automatically triggered disbursements to affected communities when a disaster struck; and (4) Actively involving local communities in disaster planning and response efforts.

The goal of adaptive social protection is to ensure that no one is left behind during times of crisis by utilizing adaptable techniques that can adapt to changing requirements and dangers (Bowen et al. 2020) ASP entails responsive measures that shield vulnerable people, boost their resilience, and give them the tools they need to recover from disasters successfully. The foundation of ASP is resilience, which enables people and communities to adapt and prosper in the face of challenges (USAID 2022). These interventions cover a wide array of actions, such as social safety nets, cash transfers, insurance plans, livelihood support, educational, medical, and employment initiatives (Alderman and Yemtsov 2012). The goal is to lessen the negative effects of catastrophes and promote sustainable recovery that includes risk reduction improve resilience and enabling communities to recover from disasters stronger (WFP 2023).

ASP focuses on identifying and addressing vulnerabilities that exacerbate the impact of disasters on marginalized groups, such as women, children, the elderly, and persons with disabilities (ASEAN 2021). Tailored programs ensure their unique needs are met. Inclusive recovery involves active engagement and participation of affected communities in the planning, design, and implementation of recovery initiatives (Trujillo 2020). Risk transfer mechanisms, such as insurance, provide a safety net for communities in the aftermath of disasters, offering financial protection and enabling a quicker and more resilient recovery (Nations 1961).

However, post-disaster recovery in developing countries presents a unique set of challenges, often characterized by limited resources, weak infrastructure, and a high vulnerability to recurring disasters (Mendis et al. 2023). These nations face an arduous task of not only rebuilding physical infrastructure but also rehabilitating communities, restoring livelihoods, and enhancing resilience to mitigate future risks (United Nations 2020). The recovery process necessitates a holistic approach encompassing various sectors and involving both national and international stakeholders.

One of the significant challenges faced in post-disaster recovery is the immense strain on limited resources (Lamb 2020). Developing countries often grapple with constrained budgets and lack the financial means to fund comprehensive recovery efforts. This financial gap impedes the ability to rebuild and rehabilitate swiftly, slowing down the recovery process. External aid and investment are crucial in bridging this gap and facilitating a more robust recovery. International organizations, donor countries, and non-governmental organizations (NGOs) play a pivotal role in providing the necessary financial assistance and expertise to aid in the recovery process (Khaledi et al. 2023).

In addition to financial constraints, weak infrastructure exacerbates the impact of disasters. Many of these nations have fragile infrastructural systems that are ill-equipped to withstand the forces of a disaster (FEMA et al. 2021). Post-disaster recovery, therefore, calls for not only restoring but also upgrading infrastructure to be more resilient and capable of withstanding future disasters. Investments in resilient infrastructure, incorporating disaster risk reduction measures, are essential for a sustainable recovery (ESCAP 2021). Furthermore, the recovery process should address the socio-economic aspects of affected communities. Disasters often disproportionately affect vulnerable populations, including the poor, marginalized, and disadvantaged (Hallegatte et al. 2020). The recovery efforts must ensure inclusivity, focusing on these vulnerable groups to mitigate the widening socio-economic gap. Targeted social protection programs, skill development initiatives, and livelihood support are essential components of a recovery strategy that addresses the needs of the most affected (ILO 2020).

The integration of Adaptive Social Protection with the Intersectional Framework represents a significant leap forward in achieving a more inclusive, adaptive, and equitable approach to disaster management (Hankivsky 2012). By recognizing and addressing the diverse needs and vulnerabilities that arise from intersecting identities, these frameworks lay the foundation for a comprehensive and effective response to the complex challenges posed by disasters. This collaborative synergy is essential for building resilient communities that can withstand and recover from the impacts of disasters while promoting social justice and equality. Both concepts underscore the significance of inclusivity, resilience, and adaptability in addressing the multifaceted challenges posed by disasters (Bello et al. 2021).

The Intersectional Framework, rooted in critical social theory, emphasizes the intersecting identities and social categories that shape individuals' experiences, vulnerabilities, and access to resources (Collins et al. 2021). It recognizes that individuals do not face singular forms of oppression but rather a complex interplay of various factors such as gender, age, socio-economic status, and ethnicity. In the context of disaster management, the Intersectional Framework highlights the diverse needs and challenges faced by different groups within a population (Kuran et al. 2020). It sheds light on the unique vulnerabilities experienced by, for instance, elderly women from marginalized communities during and after disasters. This nuanced understanding is crucial for developing targeted and inclusive interventions (Bergenek 2023) (Fig. 10.1).

ASP and the Intersectional Framework are pivotal for creating a disaster management approach that leaves no one behind. ASP's adaptability aligns seamlessly with the Intersectional Framework's emphasis on acknowledging and addressing the unique challenges faced by various demographic groups. First, ASP, guided by the Intersectional Framework, ensures that social protection programs are designed inclusively, considering the specific needs and vulnerabilities of different intersecting identities (Chaplin et al. 2019). For example, evacuation plans must consider the mobility challenges faced by differently abled individuals, and relief distributions should account for dietary restrictions based on cultural and religious practices.

Second, The Intersectional Framework urges the collection and analysis of disaggregated data, providing a nuanced understanding of diverse vulnerabilities (Hankivsky 2012). ASP utilizes this data to tailor social protection measures dynamically, adapting to the evolving needs of various groups affected by disasters. Third, ASP's financial component aligns with the Intersectional Framework by ensuring equitable resource allocation based on the differentiated needs arising from intersecting identities (UNPRPD UN Women 2020). Adequate financing is essential to implement targeted interventions and address the specific vulnerabilities of marginalized groups.

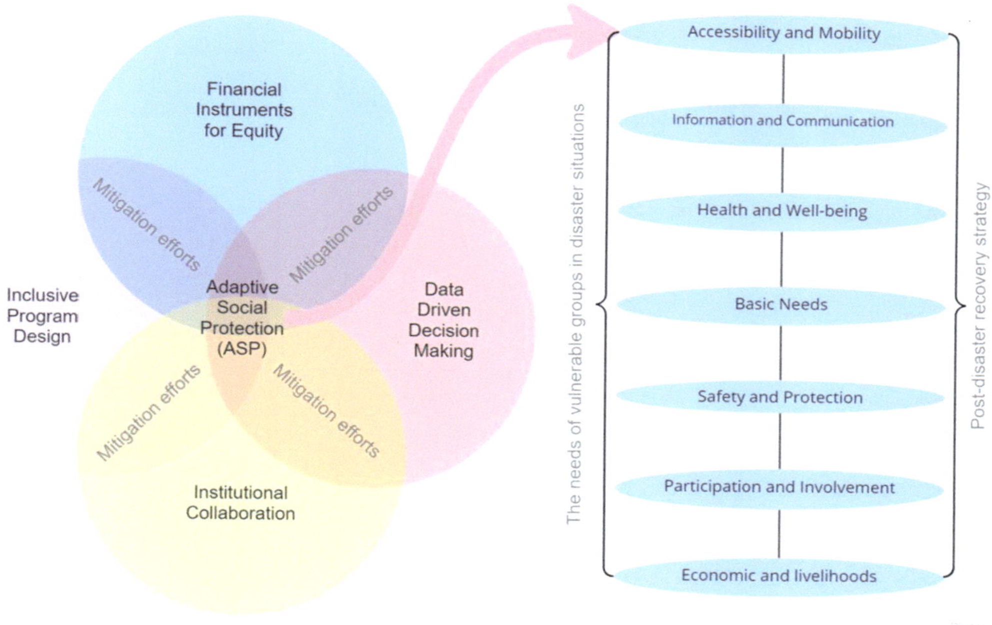

Fig. 10.1 Model for ASP disaster management approach (author)

Fourth, both ASP and the Intersectional Framework call for collaborative institutional arrangements involving governments, NGOs, and local communities (Magnusdottir and Kronsell 2021). Such collaboration ensures that diverse voices are heard in decision-making processes and that interventions are contextually relevant.

10.3 Ensuring Resilience and Inclusivity

Law Number 23 of 2014 concerning Regional Governance and Government Regulation Number 2 of 2018 regarding Minimal Service Standards has explicitly delegated disaster management authority to regional governments, aiming to establish minimum service standards in the regions, including funding for community involvement in disaster management. Furthermore, in accordance with the prioritization of village fund usage established through the Ministry of Village Regulation, village funds can also serve as supplementary funding sources in post-disaster recovery in disaster-affected villages (Mubarak et al. 2023).

The role of rural communities becomes increasingly significant in decision-making, especially within the framework of community-based disaster management at the village level. Post-disaster recovery should prioritize enhancing community capacity to support community-based post-disaster recovery (EMA 2003), including supporting disaster-prepared villages and empowering rural communities to reduce disaster risks. Here is the table detailing post-disaster recovery needs for various vulnerable groups (Table 10.1):

This table outlines the specific needs of various vulnerable groups during post-disaster recovery, aiming to ensure that they receive appropriate and effective support throughout the recovery process. The ASP framework, combined with an Intersectional perspective, provides a comprehensive approach to addressing the unique needs of various vulnerable groups.

Table 10.1 Post-disaster recovery needs for various vulnerable groups

No.	Vulnerable group	Post-disaster recovery needs
1.	Older people	Access to continuous healthcare services and medical care for chronic conditions
		Mobility assistance and transportation to access essential facilities
		Psychosocial services to cope with trauma and stress
		Safe and accessible housing
		Provision of food and nutrition that meets their health requirements
2.	Women	Access to maternity healthcare services, including prenatal and postnatal care
		Transportation assistance to access clinics and hospitals
		Adequate nutrition and dietary supplements
		Safe and comfortable housing with adequate sanitation facilities
		Psychosocial support to manage stress and anxiety during pregnancy
3.	Children	Access to education and temporary learning facilities
		Psychosocial support to cope with trauma and provide a sense of security
		Provision of nutritious food and clean water
		Safe play spaces and environments that support child development
		Protection from exploitation, violence, and trafficking
4.	Persons with Disabilities	Access to specialized healthcare services and assistive devices such as wheelchairs, hearing aids, etc.
		Accessible infrastructure including roads, buildings, and evacuation centers
		Accessible information in suitable formats (e.g., Braille, sign language)

(continued)

Table 10.1 (continued)

No.	Vulnerable group	Post-disaster recovery needs
		Psychosocial support and rehabilitation services
		Safe and accessible housing with appropriate facilities
5.	Other marginalized groups	Equitable and fair access to healthcare services
		Financial and livelihood support to restore their economic stability
		Participation in recovery planning and decision-making processes
		Provision of safe and adequate temporary housing
		Access to education and skills training to enhance self-sufficiency

10.4 The Needs of the Older People in Post-disaster Recovery

According to Law Number 13 of 1998 concerning Elderly Welfare, Article 1, Paragraph 1 states that welfare is a way of life and social livelihood, encompassing both material and spiritual aspects, characterized by a sense of safety, morality, and inner tranquility that enables every citizen to fulfill their physical, spiritual, and social needs to the best of their ability for themselves, their families, and the community, while upholding the fundamental rights and duties of humans in accordance with Indonesia constitution.

The physical aspect involves meeting the essential physical requirements necessary for sustaining life. It encompasses the need to maintain life and endure. Several unique challenges faced by the elderly in post-disaster conditions include physical weakness and helplessness, leading to dependence on others (Pujiatun et al. 2023). Their economic situation is highly threatened, necessitating new productive activities to fill the increasing leisure time. The required physical needs include sufficient and healthy food, clothing and accessories, shelter/housing, health care and supervision, practical daily technical assistance/legal aid, public transportation for the elderly, visits/companionship, healthy recreational and entertainment, a sense of security and tranquility, and assistance with sensory aids such as eyeglasses and hearing aids (Bashawri et al. 2014). Given the diverse problems and needs, the elderly need to maintain the quality of life throughout their remaining years. Older individuals are particularly vulnerable during disasters. When a disaster makes it difficult or impossible to access healthcare services, the elderly suffering from chronic illnesses will be more severely impacted than others (Maltais 2019).

The psychological aspect pertains to mental well-being that influences both physical conditions and the quality of life. When an individual is psychologically, socially, and emotionally well, it can be said that their psychological needs are fulfilled. The psychological needs of the elderly depict the conditions of their life, encompassing a sense of safety, morality, and inner tranquility, allowing every elderly person to meet their physical, spiritual, and social needs to the best of their ability for themselves, their families, and society (Arabzadeh 2017). It requires courage and a renewed spirit to make significant lifestyle changes, such as seeking new companionship after the loss of a spouse. To achieve a satisfactory quality of life, continuous protection and services are necessary so that the elderly can enjoy a reasonable standard of living (Guida et al. 2022).

Women who lose their husbands, lose their children, or daughters who lose their parents or family may experience relatively prolonged psychological trauma. The recovery of psychological trauma is crucial, with a specific approach tailored to each type of case that arises.

It is crucial for the elderly to maintain social relationships as part of disaster preparedness (Wakui et al. 2017). Disaster preparedness involves not only stockpiling resources such as food, medications, and water but also paying attention to social connections. Researchers have found that older individuals who can sustain social relationships, such as informal interactions with neighbors and family, are shielded from certain negative impacts, such as depressive symptoms (Gan and Best 2021). Those who experience depression or traumatic stress during disasters, such as the challenges of evacuation, are likely to face limitations in performing daily activities six years later. The Republic of Indonesian government has implemented programs in the field of social welfare for the elderly to maintain and improve their quality of life. These programs are carried out through social services for the elderly, both institutional and non-institutional. However, not all elderly individuals have access to social services, whether institutional or non-institutional. For the elderly exposed to disasters, avoiding relocation to temporary shelters, and making efforts to maintain social connections with family and friends can protect their health (Dynes 2006).

The spiritual aspect can assist the elderly in rejuvenating their zest for life. In times of illness or the loss of a spouse or family member during a disaster, spiritual strength can help heal emotional wounds or instill the awareness that there is still a life worth fighting for tomorrow (Nurhidayati et al. 2020). Spiritual needs are essential for every individual, particularly for the elderly, as they constitute a vital element in shaping one's character. If someone loses their spiritual needs, there is concern that they may experience spiritual distress. Spiritual distress is a condition where an individual or group experiences or is at risk of experiencing disruptions in their beliefs or value system that provide strength, hope, and meaning in life (Fitriana and Kuncorowati 2021). Therefore, the spiritual needs of the elderly must be managed and enhanced by providing motivation regarding the existence of the Almighty. The government plays a role in raising awareness among the elderly to accept their circumstances and be good listeners so that the elderly have hope of avoiding spiritual distress. The high spiritual needs of the elderly are attributed to their aging bodies and mature thoughts about facing death, leading many elderly individuals to draw closer to the Divine. Understanding the interconnected factors is crucial in designing effective interventions to enhance the health and well-being of the elderly experiencing disasters (Sandifer and Walker 2018).

10.5 Women and Children in Post-disaster Situations

The victims of the 2004 Aceh tsunami, approximately 70–80%, were women (Rahiem et al. 2021). Women often become victims in disaster situations due to their lower access to resources such as toilet facilities and clean water during evacuations (Hamidazada et al. 2019). The lack of infrastructure complicates meeting women's needs related to menstruation, pregnancy, childbirth, and breastfeeding. In evacuation centers, limited privacy spaces make many women feel uncomfortable. Insufficient privacy in shelters can lead to the emergence of sexual violence against women. Sexual violence against women even becomes a looming threat in post-disaster environments. Young girls may feel helpless in evacuation centers as they are separated from their families who could protect them (Hartono 2021).

Adolescent girls constitute a community vulnerable to sexual violence, rape that may lead to sexually transmitted infections, and the risk of unwanted pregnancies. The government must commit to addressing women as a vulnerable group in disasters. The central government, local government, non-governmental organizations (NGOs), and the community must commit to implementing disaster mitigation programs and providing continuous education to communities in disaster-prone areas. These findings emphasize the need to direct disaster preparedness and response resources toward this vulnerable group (Wiest et al. 1994).

The Learning Circle for Women (Libu Perempuan), which established women-friendly tents in post-earthquake and tsunami evacuation locations in Central Sulawesi in September 2018, received several reports of sexual violence against women. For instance, makeshift toilet and emergency bathroom facilities, the distance of toilets and bathrooms from evacuation points, and inadequate lighting also pose risks to the safety of women. A young girl, a recent earthquake victim in Central Sulawesi who sought refuge in Makassar, was allegedly raped by three men in the evacuation center.

The lack of clean water can be problematic for women, affecting not only personal hygiene but also the quantity of breast milk for breastfeeding mothers. Pregnant women, already a vulnerable population in normal conditions, become even more susceptible in post-disaster situations with an unsupportive environment for their basic needs (Tambunan et al. 2024). Crowded tents, separation from family, and various stressors can threaten the nutritional needs of pregnant women. Insufficient intake of nutritious food tailored to their nutritional requirements, along with inadequate clothing quality, are significant reasons for the vulnerability of women in evacuation. Therefore, special considerations are required for pregnant women, both in the evacuation process and in evacuation centers (Fletcher et al. 2005).

However, the lessons from the 2010 Mount Merapi eruption indicate that women have been more proactive and responsive than men (Mutiarni et al. 2022). Women have proven to be effective in disaster management, both in prevention and emergency response. The Mount Merapi eruption transformed women into resilient individuals capable of facing challenging and critical conditions (Lily Gardener 2015). Disasters have added to the hardship of these resilient women, yet they do not lose hope. Despite the challenging circumstances, they contribute by assisting in recovery programs, cooking in communal kitchens, and distributing daily provisions regularly. They do not focus on themselves but rather contribute to the recovery efforts, influencing their family, immediate neighbors, and extending to other regions (Yulianti 2021). They seek breakthroughs for a self-sustained life, aiming not to depend on the assistance and charity of others. For these resilient women, living independently through their own efforts is more fulfilling than receiving aid.

Empowering women becomes a driving force (agent of change) capable of responding to environmental changes. Resilient women do not differentiate between domestic and public work, indicating an absence of gender issues in their work (Kusumasari 2015). After the Mount Merapi eruption, disaster management programs also involved women in the recovery process. Communal kitchens, often associated with the domestication of women, reveal the true existence of women. Communal kitchen activities constitute public services that require women's organizational skills, thus not solely seen as domestic activities. Communal kitchens, for women's existence, not only imply their role in post-disaster recovery but also establish a distinctive concept of women's organizational skills.

10.6 Conclusion

Adaptive Social Protection (ASP) has been recognized as an effective policy to reduce the impact of shocks and stress triggers on vulnerable households. ASP is considered a transformative intervention at two different levels: at the system level and at the beneficiary level. At the system level, ASP programs contribute to building resilience beyond short-term mitigation strategies, which are typically the focus of humanitarian interventions. At the beneficiary level, through its activities, ASP strengthens the adaptive capacity of vulnerable households There are at least five steps needed to strengthen ASP in Indonesia, (1) the formulation of regulations and the development agenda for ASP; (2) ownership of development coordination among stakeholders; (3) risk and vulnerability analysis based on valid data and risk mapping; (4) adaptation of social protection programs according to risks and vulnerabilities; and (5) the development of sustainable financing for program implementation.

The scheme of social protection in enhancing disaster resilience and recovery includes the following: First, data transformation toward social registration through data improvement. Data transformation should focus more on three aspects: immediacy, confidence/credibility, and validity. Second, digitization of social aid distribution using an integrated digital platform, the opening of a single social aid account, and payments through Fintech. The Ministry of Social Affairs plays a crucial role in creating a system and building supporting infrastructure to realize the concept of digitizing social aid payments. Third, periodic evaluation, coordination, and discussions with relevant parties are necessary to analyze the effectiveness of social protection programs using information technology through virtual meetings. Fourth, in budgeting data improvement, in line with the national budgeting principle, it is necessary to adhere to the principle of money following the program, so the program must be clear first, what program will be implemented, and who will lead or be responsible for the program. Fifth, there is a need to develop a more adaptive social protection scheme to adapt to various disasters such as floods, landslides, earthquakes, tsunamis, and droughts. Sixth, regarding security constraints in the distribution of in-kind aid (rice, sugar, oil, etc.) to disaster-prone areas, collaboration with relevant authorities must be involved.

It is important to highlight how community resilience is shaped not only by social cohesion and adaptive capacity but also by intersecting factors such as gender, class, race, ethnicity, and power dynamics. Intersectionality enriches the analysis of disaster risk reduction (DRR) by recognizing that different groups within a community experience vulnerabilities and resilience-building processes in unique ways based on their positionalities. An intersectional framework in DRR acknowledges the diversity within communities, ensuring that responses are tailored to meet the needs of various social groups. Community resilience is not experienced uniformly; factors such as gender, socio-economic status, and cultural background influence how individuals and groups prepare for, respond to, and recover from disasters. By incorporating intersectionality and ASP, policymakers and practitioners can better understand the complex, multifaceted nature of vulnerability and resilience, ensuring that interventions are more inclusive and equitable.

References

Agustawijaya DS (2013) A review on hazard risk reduction systems and reliability estimate of the dredging system of the Lusi Mud Volcano in Sidoarjo, East Java. Int J Civ Environ Eng IJCEE-IJENS 13:2

Alderman H, Yemtsov R (2012) Productive role of safety nets. Soc Protect Labor Discuss Pap 1203:1–93

Almutairi A, Mourshed M, Ameen RFM (2020) Coastal community resilience frameworks for disaster risk management. Nat Hazards. https://doi.org/10.1007/s11069-020-03875-3

Arabzadeh M (2017) Relationship between basic psychological needs and mental vitality in the elderly. Salmand 12(2):170–179. https://doi.org/10.21859/sija-1202170

ASEAN (2021) ASEAN guidelines on disaster responsive social protection to increase resilience. In: Asean

Bashawri A, Garrity S, Moodley K (2014) An overview of the design of disaster relief shelters. Proc Econ Financ 18(September):924–931. https://doi.org/10.1016/s2212-5671(14)01019-3

Behera S, Das A (2023) Methodology for the post-disaster damages, losses, and need assessment: a review, p 12

Bello O, Bustamante A, Pizarro P (2021) Planning for disaster risk reduction within the framework of the 2030 Agenda for sustainable development. Project Documents, LC/TS.2020, pp 1–61

Bergenek VCJ (2023) Vulnerability and resilience in an ageing population, May

Bollin C, Khanna S (2007) Review of post disaster recovery needs assessment and methodologies—experiences from Asia and Latin America, November, p 110

Bowen T, del Ninno C, Andrews C, Coll-Black S, Gentilini U, Johnson K, Kawasoe Y, Kryeziu A, Maher B, Williams A (2020) Adaptive social protection: building resilience to shocks. In: Adaptive social protection: building resilience to shocks. https://doi.org/10.1596/978-1-4648-1575-1

Cazabat C, Anzellini V, Fung V, Housset T, O'Connor A, Ponserre S (2023) Disaster displacement: Indonesia country briefing, pp 1–28

Chaplin D, Twigg J, Lovell E (2019) Intersectional approaches to vulnerability reduction and resilience-building. Resilience Intel 12:35. https://doi.org/10.13140/RG.2.2.13404.03209

Collins PH, da Silva ECG, Ergun E, Furseth I, Bond KD, Martínez-Palacios J (2021. Intersectionality as critical social theory: intersectionality as critical social theory, Patricia Hill Collins, Duke University Press, 2019. Contemp Polit Theory 20(3):690–725. https://doi.org/10.1057/s41296-021-00490-0

Dynes RR (2006) Naval postgraduate school homeland security affairs (Issue 2). http://hdl.handle.net/10945/25095

Elnashai AS (2007) Mid-America earthquake center 1

EMA (2003) Community development in

ESCAP (2021) Climate and disaster resilient infrastructure: building resilience to future uncertainties and shocks, pp 1–60

Fakhruddin I, Elmada MAG (2022) Local wisdom as a part of disaster communication: a study on the local storytelling in disaster mitigation. ETNOSIA J Etnografi Indonesia 7(2):154–166. https://doi.org/10.31947/etnosia.v7i2.22145

FEMA, Bello O, Bustamante A, Pizarro P, Johnson K (2021) Disaster technical assistance center supplemental research bulletin greater impact: how disasters affect people of low socioeconomic status. Project Documents, LC/TS.2020(September), pp 1–96

Fitriana KN, Kuncorowati PW (2021) Disaster mitigation for elderly: the challenges of social welfare services for elderly in the COVID-19 pandemic. IOP Conf Ser Earth Environ Sci. https://doi.org/10.1088/1755-1315/884/1/012048

Fitrinitia IS, Matsuyuki M, Nakamura F, Tnaka S, Ariyoshi R (2022) An integration of social protection, disaster risk reduction, and climate change adaptation in disaster policy, plan and program: a review on global discourse and case study of Indonesia. J Jpn Soc Civ Eng Ser D3 (Infrastruct Plan Manag), 78(6), II_302-II_314. https://doi.org/10.2208/jscejipm.78.6_ii_302

Fletcher L, Stover E, Weinstein H, Org E (2005) UC Berkeley reports title after the Tsunami: human rights of vulnerable populations permalink. https://escholarship.org/uc/item/7g73x8mj. Publication Date. https://escholarship.org/uc/item/7g73x8mj

Gan DRY, Best JR (2021) Prior social contact and mental health trajectories during covid-19: neighborhood friendship protects vulnerable older adults. Int J Environ Res Public Health. https://doi.org/10.3390/ijerph18199999

Gasior K, Wright G, Barnes H, Noble M (2023) Adaptive social protection in Indonesia—stress-testing the effect of a natural disaster on poverty and vulnerability. CeMPA working paper series, 01/23(January)

Guida C, Carpentieri G, Masoumi H (2022) Measuring spatial accessibility to urban services for older adults: an application to healthcare facilities in Milan. Eur Transp Res Rev. https://doi.org/10.1186/s12544-022-00544-3

Gunawan I, Sopaheluwakan J, Sagala SAH, Zawani H, Amin S, Mangunsong RTJ (2016) Building Indonesia's resilience to disaster: experiences from mainstreaming disaster risk reduction in Indonesia program

Hagan AB, Foster R, Perera N, Aji C (2007) Tsunamis and coral reefs

Hallegatte S, Vogt-Schilb A, Rozenberg J, Bangalore M, Beaudet C (2020) From poverty to disaster and back: a review of the literature. Econ Dis Clim Change 4(1):223–247. https://doi.org/10.1007/s41885-020-00060-5

Hamidazada M, Cruz AM, Yokomatsu M (2019) Vulnerability factors of afghan rural women to disasters. Int J Dis Risk Sci 10(4):573–590. https://doi.org/10.1007/s13753-019-00227-z

Hankivsky O (2012) An intersectionality-based policy analysis framework. In: Institute for intersectionality research and policy

Hartono Y (2021) Gender-based sustainable disaster response education to anticipate the problem of women's vulnerability. http://sjdgge.ppj.unp.ac.id/index.php/Sjdgge

ILO (2020) Employment, livelihoods and social protection guide for recovery implementation

Jibiki Y, Pelupessy D, Sasaki D, Iuchi K (2020) Implementation of post disaster needs assessment in Indonesia: literature review. J Dis Res 15(7):975–980. https://doi.org/10.20965/jdr.2020.p0975

Khaledi H, Marzaleh MA, Peyravi M, Rezaee R (2023) Non-governmental organizations (NGOs) in disaster management: a qualitative study. Trauma Monthly 28(1):678–686. https://doi.org/10.30491/tm.2023.368785.1525

Kuran CHA, Morsut C, Kruke BI, Krüger M, Segnestam L, Orru K, Nævestad TO, Airola M, Keränen J, Gabel F, Hansson S, Torpan S (2020) Vulnerability and vulnerable groups from an intersectionality perspective. Int J Dis Risk Reduct. https://doi.org/10.1016/j.ijdrr.2020.101826

Kusumasari B (2015) Women adaptive capacity in post disaster recovery in Indonesia. Asian Soc Sci 11(12):281–289. https://doi.org/10.5539/ass.v11n12p281

Lamb Z (2020) Louisiana's response to extreme weather: a coastal state's adaptation challenges and successes

Lily Gardener J (2015) Under the ash cloud. Gender distinctions in the Resilience of Kemiri Community towards Mount Merapi volcanic activity. IFHV working paper, 5(1). https://doi.org/10.17176/20220622-140257-0

Magnusdottir GL, Kronsell A (2021) Gender, intersectionality and climate institutions in industrialised states. In: Gender, intersectionality and climate institutions in industrialised states. https://doi.org/10.4324/9781003052821

Maltais D (2019) Elderly people with disabilities and natural disasters: vulnerability of seniors and post trauma. Gerontol Geriatr Med 5(4):1–7. https://doi.org/10.24966/ggm-8662/100041

Mei ETW, Fajarwati A, Hasanati S, Sari IM (2016) Resettlement Following the 2010 Merapi Volcano Eruption. Proc Soc Behav Sci 227(November 2015):361–369. https://doi.org/10.1016/j.sbspro.2016.06.083

Mendis K, Thayaparan M, Kaluarachchi Y, Pathirage C (2023) Challenges faced by marginalized communities in a post-disaster context: a systematic review of the literature. Sustainability (Switzerland). https://doi.org/10.3390/su151410754

Mubarak MR, Nurhalizah D, Samintang S, Rahadian A, Ampri AII (2023) Index-based insurance through village fund allocation as a constructive solution towards post-disaster agricultural recovery for vulnerable farmers

Mutiarni YS, Nakamura H, Bhattacharya Y (2022) The resilience community: strengthening people-centered disaster risk reduction in the Merapi Volcano Community, Java, Indonesia. Sustainability (Switzerland). https://doi.org/10.3390/su14042215

Nations U (1961) Economic and social council. Int Organ 15(3):500–510

Nefianto T (2023) Policy and public management on earthquakes. JPPI Jurnal Penelitian Pendidikan Indonesia 9(2):357–362. https://doi.org/10.29210/020232248

Nurhidayati I, Agustiningrum R, Aprilia Ningtyas D (2020) Elderly's disaster resilience in natural disaster: literature review

Nurwihastuti DW, Sartohadi J, Mardiatno D, Nehren UR (2014) Understanding of earthquake damage pattern through geomorphological approach: a case study of 2006 Earthquake in Bantul, Yogyakarta, Indonesia. World J Eng Technol 02(03):61–70. https://doi.org/10.4236/wjet.2014.23b010

Plumlee BGS, Casadevall TJ, Wibowo HT, Robert J, Johnson CA, Breit GN, Lowers HA, Ruth E, Hageman PL, Goldstein H, Anthony MW, Cyrus J, Fey DL, Meeker GP, Morman SA, Survey USG (2008) Preliminary analytical results for a mud sample collected from the LUSI Mud Volcano, Sidoarjo, East Java, Indonesia

Pujiatun P, Zahra KA, Nurdiati DS (2023) Health-related quality of life of the elderly after natural disasters: a scoping review. J Med Sci (Berkala Ilmu Kedokteran) 55(2):179–194. https://doi.org/10.19106/jmedsci005502202310

Rahiem MDH, Rahim H, Ersing R (2021) Why did so many women die in the 2004 Aceh Tsunami? Child survivor accounts of the disaster. Int J Dis Risk Reduct. https://doi.org/10.1016/j.ijdrr.2021.102069

Rofi A, Doocy S, Robinson C (2006) Tsunami mortality and displacement in Aceh province, Indonesia. Disasters 30(3):340–350. https://doi.org/10.1111/j.0361-3666.2005.00324.x

Ruswandi D, Maarif S (2019) Implementation of disaster management policy in Indonesia. J Crit Rev 6(5):67–70. https://doi.org/10.22159/jcr.06.05.11

Sandifer PA, Walker AH (2018) Enhancing disaster resilience by reducing stress-associated health impacts. Front Public Health. https://doi.org/10.3389/fpubh.2018.00373

Supendi P, Nugraha AD, Widiyantoro S, Pesicek JD, Thurber CH, Abdullah CI, Daryono D, Wiyono SH, Shiddiqi HA, Rosalia S (2020) Relocated aftershocks and background seismicity in eastern Indonesia shed light on the 2018 Lombok and Palu earthquake sequences. Geophys J Int 221(3):1845–1855. https://doi.org/10.1093/gji/ggaa118

Syamsidik, Al Farizi MD, Tursina, Yulianur A, Rusydy I, Suppasri A (2023) Assessing probability of building damages due to tsunami hazards coupled with characteristics of buildings in Banda Aceh, Indonesia: a way to increase understanding of tsunami risks. Int J Dis Risk Reduct. https://doi.org/10.1016/j.ijdrr.2023.103652

Tambunan R, Siregar KN, Sipahutar T (2024) Spatial analysis using exploratory spatial data analysis (ESDA) of child marriage among women 15–24 years old in Indonesia. Med Publ Promosi Kesehatan Indonesia (MPPKI) 7(6):1642–1650. https://doi.org/10.56338/mppki.v7i6.5406

Trujillo M (2020) Guidelines for community participation in disaster recovery. Undp 1999(December):1–6

United Nations (2020) How developing countries are addressing hazards, focusing on relevant lessons learned and good practices, p 28

USAID (2022) 2022 resilience policy revision. December

Wagstaff A, Gwatkin DR, Johnson K, Rutstein S, Suliman E, Amouzou A (2019) Indonesia-community-based settlement rehabilitation and reconstruction project for central and West Java and Yogyakarta special region, p 144451

Wakui T, Agree EM, Saito T, Kai I (2017) Disaster preparedness among older Japanese adults with long-term care needs and their family caregivers. Dis Med Public Health Prep 11(1):31–38. https://doi.org/10.1017/dmp.2016.53

WFP (2023) Linking disaster risk financing to social protection in the commonwealth of Dominica (Issue October)

Wiest RE, Mocellin JSP, Thandiwe D (1994) The needs of women in disasters and emergencies

Women UNPRPDUN (2020) Intersectionality resource guide and toolkit: an intersectional approach to leave no one behind. Photonics Spect 36(5):58

Yudistira D, Fadilah RN, Setiawan AB (2020) The impact of Merapi mountain eruption on the community economy. Efficient Indones J Develop Econ 3(1):719–725. https://doi.org/10.15294/efficient.v3i1.36695

Yulianti A (2021) Public engagement and community resilience-building strategies in post-disaster redevelopment planning a case study of post-eruption Mount Merapi in Sleman regency (Indonesia)

Zhao B, Liao H, Su L (2021) Landslides triggered by the 2018 Lombok earthquake sequence, Indonesia. CATENA 207(April):105676. https://doi.org/10.1016/j.catena.2021.105676

Fixing the State? A Reflection of Civil Society Response to COVID-19 Crisis in Indonesia

11

Jonatan A. Lassa

Abstract

CSOs' and/or NGOs' behavioral traits as policy entrepreneurs, including being policy drafters as they function as queasy parliamentarians, have been widely recognized in the literature. Emerging traits less known to many include the fact that during crises where networked CSOs endeavored to 'fix the state' and 'jump-start local governance' during crises and disasters. COVID-19 offered a unique opportunity to see how uncommon behavioral traits are shared by Indonesian CSOs. This study is based on a case from the early stages of COVID-19, when the Pandemic was practically uncharted territory where no government was immune from making mistakes, including underestimating pandemic signals and delays, which reflect various cognitive biases of public leaders and policymakers. This paper examined a sub-national CSO network that demonstrated how civil society united and collectively complemented the government's lukewarm responses during the initial stage of COVID-19, from January until the end of 2020. The case demonstrates the imagination of CSO actors in their efforts to 'fix the state' and complement the governmental response. The 'fixing the state's approach includes strategically guiding state officials to make proper decisions and adopt mitigation measures, as well as building the capacity for whole-of-society coordination beyond the 'whole-of-government's state-centric approach, which proved ineffective in addressing a large-scale, multi-dimensional crisis.

Keywords

Civil society organizations (CSOs) · Policy entrepreneurs · COVID-19 response · Networked governance · State-civil society relations

11.1 Introduction

With the growing dominance of the neoliberal paradigm in public policy, the architecture of disaster management in many countries, including Indonesia, has changed with considerations for efficiency in public policy. A common slogan expressed by state officials is 'disasters are everyone's responsibility.' This paradigm deliberately capitalizes on civil society participation by making it seem like an obligation, as described in Law 24/2007, which interprets the principle of 'shared responsibilities' to mean that 'disaster management is essentially the responsibility of

J. A. Lassa (✉)
Institute of Earth Science (GNS Science), Lower Hutt, New Zealand
e-mail: jonatan.lassa@cdu.edu.au

D. Hidayati et al. (eds.), *Examining Disaster Risk Reduction in Indonesia*,
Sustainable Development Goals Series, https://doi.org/10.1007/978-981-96-4182-6_11

governments and communities to be carried out in mutual cooperation.'

Elsewhere in the world, the process of transferring government responsibilities to Civil Society Organizations (CSOs) (both individuals and collectives) occurs covertly and can be seen as a trace of the neoliberal paradigm, which increasingly demands that civil society and/or taxpayers (broadly defined as individuals, families, communities, and businesses) must manage their disaster risks either individually or collectively (Quiggin 2007). One implication is the streamlining of disaster management bureaucracy. For example, the Australian model (Eburn and Dovers 2014), where emergency services are typically kept small and disaster response capacity is expanded by involving volunteers.

Civil society participation in disaster management is a right guaranteed by the state. The Indonesian Constitution requires the state to protect its people's safety at all times. The National Disaster Law (No. 24/2007) also emphasizes peoples' rights to state protection from various threats, including different types of disasters, such as natural hazard-induced disasters. These rights include the right to disaster information, social protection, participation, oversight of disaster response implementation, and involvement in emergency disaster operations (see Article 26).

Therefore, it can be understood that the government's preference for civil society participation, including volunteers—whether organized or not, spontaneous (digital), or formal-conventional—is primarily based on the logic of formal institutional efficiency rather than a vision of channeling aspirations and upholding the rights of association and participation in the spirit of democracy and inclusiveness. Unfortunately, in the context of crises, this obligation often positions civil society as merely accepting top-down policies rather than deliberatively allowing space for civil society to meaningfully complement the formal emergency management process.

Thus, ideally, governments (and CSOs themselves) need to position CSOs in crisis and disaster response praxis as complementary forces, not as substitutes. However, the empirical reality of emergency and humanitarian response does not simply end in a binary scenario of CSOs' roles being complementary vs. substitution. As shown later in the article, in certain contexts of time and space, CSOs' responses can act beyond complementary roles to temporarily 'jump-start' the formal institutions that initially struggle or seem passive (or even clueless) in taking clear command during disaster or pandemic (COVID-19) emergency operations.

In the contemporary history of global humanitarian disaster response in the Global South, the phenomenon of CSOs jump-starting formal executive institutions often occurs in the context of disaster management coordination. For instance, when government apparatuses in a disaster-affected area also become affected and/or victims, higher levels of government can control emergency response processes. In failed state contexts, the global humanitarian actors often temporarily 'take over' or 'jump start' the emergency response command operation until the host government can regain control over the emergency and recovery efforts. This model of crisis governance is often implemented in a shadowing form, where NGOs act like shadow institutions operating off-stage (behind the scenes) and occasionally on-stage to provide services (including information and communication services) to disaster survivors and responders.

This article examines the civil society's paradigm at the initial stage of COVID-19 in early 2020, when the Indonesian government attempted to formulate policy options and craft a narrative to justify crisis management decisions. These decisions often appeared ambiguous, as the government placed its trust in the formal system—including the COVID-19 crisis management structure believed to be capable of saving two things simultaneously: the economy and people's lives from the threat of COVID-19. Initially, the dichotomy between saving the economy and saving lives was perceived as mutually exclusive. Such a government gesture above was seen as going against the basic principle of crisis management, where health (human lives) must be prioritized over the economy.

To address the public's disagreement, a new mantra called 'new habit adaptation' was introduced, which can be understood and narrated as the 'new normal.' This was then institutionalized or mainstreamed and executed at all levels of government and society in the shortest possible time. In relation to 'civil society,' the author consciously attempts to narrow down civil society to the context of civil society organizations (CSOs). In a broad sense, CSOs can refer to social units with identifiable collective characteristics generally familiar to readers. For instance, CSOs can include associations of certain communities, community organizations (whether primordial, sectarian, or non-sectarian), Non-Governmental Organizations (NGOs), intellectual organizations, and humanitarian networks at the national level and local levels, as well as organized digital volunteers like Kawal COVID-19 (kawalcovid19.id), which emulates *Kawal Pemilu* (kawalpemilu.org), or the Citizen Coalition for Reporting COVID-19 (laporcovid19.org).

This paper asks two main questions: (1) how do NGOs/CSOs act collectively in the pandemic? (2) how best can civil society create their own inclusive and collaborative 'command structure' while performing its traditional task of complementing the government works at best while fixing the states whenever necessary and possible?

11.2 Hypothesis and Methods

This paper hypothesizes that networked CSOs think like the state (thinking like a state) and are structured (and therefore have the ability to act) like a shadow of state structure, but not operating in the sense of a pseudo-government. Participant observation was chosen as a method in which the author, deliberately observes civil society networks at the subnational level. It is also fair to suggest that the authors have been part of Indonesia's CSO movement and intellectual activism over the last 20 years from a periphery (initially) and at a distance (lately). The author selected the case study based on CSO experiments from the sub-national (provincial) level. The case study was based on a mixed-method qualitative approach, combining media analysis and policy document analysis with participant observation. The CSO network being examined is Forum Academia Nusa Tenggara Timur (FAN). Broadly, this article is organized into the following sections: Findings based on case of FAN Network at NTT Province. The article will end with a final remark as a final discussion and conclusion.

11.3 Findings: Civil Society Roles in East Nusa Tenggara

11.3.1 Indonesian CSOs Evolution of Post-1998 Reformasi: Shifts from Resistance to Collaboration

In general, within the contemporary context of Indonesia, post-Reformasi 1998, particularly in Nusa Tenggara Timur (NTT) in Eastern Indonesia, networked civil society organizations (CSOs) tend to adopt and maintain a position of resistance against the state. In a detailed thesis, Karen Campbell-Nelson explained that resistance manifests as curriculum or learning regime that allowed survivors or victims to articulate their realities, including identity, interest, and position concerning power holders (state institutions or officials) (Campbell-Nelson 2003). It is a learning regime because through resistance and opposing extractive development (mining) practice, local activists and communities found and/or created the space for learning and finding solutions to managing crises. The crisis is often associated with cases such as disaster events or environmental crises resulted from lack of good environmental governance from both government and extrative industries.

The author observes that in the context of CSOs with an advocacy background and intellectual movement base in NTT, there is a tendency for CSOs to adopt a corrective approach toward the state's top-down actions. Historically (in a historical path-dependent sense), advocacy-oriented CSOs tend to view the state as an authority often characterized by excessive power exercise, top-down governance, and state-sponsored bullying with shrinking spaces for democratic dialogue that have become more evident during the COVID-19 context. Therefore, government regimes at various levels tend to start with a 'deficit account' in terms of CSOs' trust. This is not surprising because the shrinking space for inter-stakeholder dialogues is a significant indication of the decline in the quality of Indonesia's democracy during 2014–2020 (Mietzner 2020).

The author does not intend to suggest that there is only one type of CSO/NGO character, as many types of CSOs have diverse relationships with the state. With the rise of a collaborative governance framework that promotes the concept of NGOs as collaborators and collaboration as ways of working with the government, in many cases, at least based on my observation, NGOs are more incentivized to work with (local) governments. Many critical networked CSOs became disincentivized if they prefer pathways to change through resistance. Besides, as a young democratic country, many local leaders are not convenient enough to share the views that critics of CSOs are basically the most efficient way of governing—as monitoring and quality control services are provided freely.

As a region with the highest democratic participation index in Indonesia (Said 2019) over the past few decades, NTT has pockets of advocacy-oriented CSOs that are vocal against the state's tendency toward predatory behavior and corruption. Building trust becomes extremely challenging in a context where the central and provincial government regimes are entering a new period of leadership filled with turbulence following the 2018 presidential election. In NTT, the provincial government is experiencing various dynamics related to agrarian issues and vertical conflicts between the state and civil society, where, in extreme cases, the government tends to use fear and terror-based politics as mechanisms of governing by fear. Unsurprisingly, the state's predatory character in agrarian issues, particularly regarding customary land versus public and private investment interests, is responded to by some as a form of 'state terrorism' (see Tan 2020).

CSO's response tends to be independent of state policy and is not merely a knee-jerk response to a governance deficit staggered by COVID-19. In this regard, the character of CSOs in crisis response was highly diverse and (though not always, but often) determined by the paradigm and the form of or structure of the civil society organization itself (endogenous factors). Examples of endogenous factors can be seen, though not limited to, the phenomenon of CSOs that have two main characteristics: namely, CSOs that think like the state (thinking like a state) and are structured (and therefore can act) like a shadow state, but not in the sense of a pseudo-government.

Symbolic and semiotic responses (Danusiri 2020) were also embedded in the government's approach, which tends to use strategic ambiguity (Lassa and Booth 2020). The author hypothesizes that, over the past 30 years, the Indonesian government has tended to be reactive when facing disaster or crisis due to the nature of high uncertainty, complexity, and ambiguity in large-scale crises and/or disasters. Due to the reactive nature toward disasters, the state tends to adopt symbolic and semiotic response strategies in times of catastrophe. Examples of such responses include the creation of terminologies and abbreviations such as the implementation of large-scale social restrictions (PSBB) and 'New Normal' (NN) or 'adapting to new behavior' (AKB). When facing public resistance, the government consciously creates ambiguous, multi-interpretive terminology, followed by rapid revisions, as seen in the '*mudik/pulang kampung*' (going home for the holiday) terminology in April/May and NN/AKB in June–July 2020. Governing by 'confuse if you can't convince' had been sometimes a strategy in the early days of COVID-19.

This paper argues that there was an emerging form of counter-semiotic or counter-symbolic response consciously or unconsciously adopted by CSOs. It should be noted, however, that this counter-semiotic process is not conducted in direct opposition but rather as resistance using a similar strategy: both the government's semiotics and the counter-semiotics of CSOs are part of a strategy or stages where every party created their learning opportunities.

Before further mapping the praxis of CSO responses and their complementary role, we need to address the question of what underlies the structure or order of disaster management in a country.

So, how about Indonesia in general? It is not surprising if this question is answered in various ways. Even within the paradigm of community-based disaster management, disaster risk reduction (DRR) activists often think ambiguously. For instance, some argue that all paths to disaster resilience must be community-based, but what is envisioned is merely the village as one administrative unit tasked with—or more precisely, managing its community members' resilience (Lassa et al. 2018). Some might argue that Indonesia's basic unit of governing power of disaster response, at least since 2008, is the local disaster management agency (or BPBD: *Badan Penanggulangan Bencana Daerah*).

Crisis management bureaucrats from BPBDs might promptly respond to natural hazard-induced disasters. COVID-19 was initially understood as a public health crisis. However, as its impact became multi-dimensional and required cross-institutional and cross-sectoral actions, President Joko Widodo decided to appoint the Head of the National Disaster Management Agency (BNPB) as the Chair of the National COVID-19 Task Force (Djalante et al. 2020; Widaningrum and Mas'udi 2020). Although there was some resistance to this concept from the public health profession, there was some justification for it. Over the past 15 years, Indonesia's health system had not been routinely tested by major crises, while the disaster management 'sector' in the form of Bakornas/ Satlak, the BNPB/BPBD predecessors, had more frequently faced large-scale disasters over the last 30 years.

At the sub-national (provincial) level, the by-default preference for the head of a local crisis management Task Force was the head of the BPBD and the Regional Secretary as the ex-officio head of the BPBD (BNPB 2008). Unfortunately, in the NTT context, CSOs and local media tend to perceive that the leaders of the BPBDs across the province appeared unable to take on leadership roles in responding to a crisis of pandemic scale. This was evident in many provinces, and the author attempts to use the case of East Nusa Tenggara as an example.

Within the global health system adopted by Indonesia, the Six Building Blocks of health system development are also recognized, such as health services, the health workforce, information systems, access to basic medical treatments, financing systems, and leadership and governance (WHO 2010). Generally, the COVID-19 crisis has exposed the weaknesses in health service investment, where sub-national laboratory systems in the regions were unprepared to be the basic infrastructure for epidemiological monitoring and control.

11.3.2 Government Response: Semiotic Strategy, Counter Resistance and Learning Curriculum

I have observed that the Indonesian government, over the past 30 years, has tended to adopt symbolic responses and various semiotics formation and creation as the primary arena or locus of response when facing uncertainty and the complexity of crises (Lassa 2020). Thus, the state has made the semiotic response. This is partly to buy in time to learn and respond to both the crisis and the pressure from the public. In other words, the state adopted a semiotic strategy as a learning regime (that is, how the state understands crises—at a particular stage, it could be called a narrative response, although without an adequate amount of meaningful or thoughtful narrative) simultaneously serves as a crisis management and leadership strategy.

Understanding semiotic responses as a learning regime is a heuristic process—a mental shortcut—a strategy employed by those who felt inferior and overwhelmed by uncertainty, often positioning themselves as victims within complex power dynamics. In this context, government officials as state representatives may feel inferior and uncertain when faced with natural or non-natural threats that are perceived to exceed the limits of formal institutional resilience.

When the ruling government feels existentially threatened, it may deem it necessary to create new symbols, terms, or meanings. This occurs when the regime finds it challenging to understand complex situations and take control over existential threats posed by not only by disasters, crises, or catastrophes, but also by, political opposition and CSOs. It is termed existential because crises of the scale of COVID-19 could create conditions where political stability could be at stake. It is important to emphasize that over the past 12 years, significant progress has been made in disaster management in Indonesia from an institutional perspective. Establishing nearly 500 new institutions in the form of Regional Disaster Management Agencies (BPBD) serves as tangible evidence.

However, this institutional progress does not eliminate the character of the state in making semiotics a crucial domain in disaster crisis response. Therefore, the strategic functions of symbolic response as a 'learning curriculum' has become a by default mental framework for Indonesian crisis management.

11.3.3 CSOs and Counter-Semiotics: The Case Study of Forum Academia NTT

This 'ethnographic' case study briefly explaining this phenomenon is the network of CSOs in Eastern Indonesia, specifically the praxis demonstrated by Forum Academia NTT (FAN). This forum is an intellectual network of NTT scholars and activists (or scholar-activist) that evolved from a mailing-list group since 2004, whose members are spread across various parts of the world, able to demonstrate collective action in assisting the government and the public in the context of coordination and communication. The author has been part of the core management of this forum from the beginning and has served as a participant observer during the COVID-19 period from a distance.

From March to July 2020, FAN took on a role in helping coordinate the COVID-19 response. The author summarizes the COVID-19 management model using a risk management (R) framework based on the triple-helix model, namely Risk = exposure × hazard × vulnerability. Managing the hazard in the context of COVID-19 was understood as the process of disinfection and identifying the COVID-19 threat through testing—using valid methods such as swab tests. However, as one of the poorest provinces in Indonesia, NTT did not have an adequate biosafety laboratory system. It took almost three months for NTT to be able to conduct testing in the city of Kupang, with a limited number of tests performed per day.

As of now most people practically know that managing exposure could be achieved through physical distancing, contact tracing, and using masks. Meanwhile, reducing vulnerability could be done by improving immunity through nutrition and exercise, which could be implemented as a long-term public policy. When the concept of the New Normal was introduced and adopted by the regional government in the mid-2020, the people in NTT seemed to merely return to the old normal—at least in the view of FAN activists and church organizations such as the Christian Evangelical Church in Timor (GMIT).

As Li (2020b) put it, 'We think as citizens from the periphery and decided to do something' in the context of the fight against COVID-19, where the state seemed shaky. 'The people have the opportunity to determine the direction of the state in times of crisis.' Therefore, efforts to support national resilience do not necessarily have to rely on formal institutions, including the legislature, which seemed passive with minimal breakthroughs. Some FAN activists saw the New Normal as an opportunity to respond with breakthroughs and innovations for mass testing, with adequate accuracy and lower costs. One solution was to propose pool testing with the qPCR method. The qPCR approach was eventually accepted by the provincial government of NTT and the City Government of Kupang in early July 2020 through various advocacy approaches.

Despite support from the central government and foreign aid, and the experience of responding to national-scale disasters over the past 30 years (since the Flores Earthquake of 1992—National

Disaster during the New Order era) to international scale (since handling East Timor refugees in West Timor 1999–2000), which various foreign aid facilities followed, the local government was not automatically capable of implementing proactive measures. In the view of some prominent activists, formally, the local government still struggled in disaster response (Li 2020b). COVID-19 exposed the quality of local administrators, who lacked independent thinking and action during critical situations (Li 2020a; b).

The good news is that civil society—at least during the COVID-19 era—appeared more capable of organizing and pushing state elements to make actual breakthroughs in innovation. From March to June 2020, FAN moderators formed a media team that later transformed into a kind of cross-institutional and cross-district coordination forum in the form of regular platforms via 'zoominars' with practical agendas such as inter-district coordination on major islands in NTT, like Flores, Sumba, and Timor. An exchange of information process then occurred, where island districts such as Rote-Ndao, Sikka, and others shared experiences in creating health processes and protocols, monitoring, and border management.

The praxis of CSOs like FAN represents a form of resistance that is converted into processes of steering formal institutions toward accountability while upholding the responsibilities demanded by the constitution and disaster management legislation. It is a process of 'fixing the state' from the periphery.

11.4 Final Remarks

In this context, complementarity and substitution become a kind of continuum where Civil Society Organizations (CSOs) can steer the leadership of executive agencies. The author's observations on how coordination meetings facilitated by non-state actors, both local and international, took place after the 2015 Nepal earthquake show that global CSOs, such as Catholic Relief Services, became key actors guiding the recovery processes in the housing sector (Magar 2019).

In Central Sulawesi, there was additional evidence of this during the early stages of the 2018 Palu earthquake. The COVID-19 event and its management dynamics also provided space for civil society to play various roles within and beyond the continuum of complementary versus substitution of state roles. The humanitarian imperative allowed CSOs to take crisis management actions based on a 'no-regret' approach, regardless of the quality and intensity of crisis management by the central and local governments.

Official data from COVID19.go.id indicates a crisis of legitimacy for the government in convincing civil society regarding the effectiveness of the 'adaptation of new habits' norm. However, civil society, which also developed COVID-19 monitoring, such as LaporCovid19 (laporcovid19.org), consistently questioned the validity of government data, which seemed less transparent, and the potential for 'score manipulation' of COVID-19 deaths and cases (Arief 2020). Examples of similar networked CSOs at the national level, such as SEJAJAR emerged as a Strategic Alternative Information Hub. SEJAJAR is a network of networks of CSOs/NGOs, including Red Cross network, that served as a platform for multi-sectors and multi-issues as long as the COVID-19 pandemic continues. As of December 2020, SEJAJAR's internal hierarchy, as comprised of 25 national organizational networks, bringing at least 600 CSOs/NGOs from 34 provinces to be a platform through which information and knowledge around COVID-19 mitigation was transmitted—some of the beneficiaries were local government officials, include more than 2000 local health workers nationwide.

When state-centric paradigm dominates the crisis information, grassroots information is needed to create more balanced information (Lassa et al. 2020). For example, with the introduction of a 'new normal' policy to relax measures of crisis management, the Indonesian government came up with a more restricted information policy. As a result, it provided less public information about COVID-19. Adequate studies suggest that a lack of information can trigger a greater sense of uncertainty and insecurity. SEJAJAR created alternative channels of information exchanges by cutting the lengthy bureaucratic procedures of vertical crisis and risk communication flows to the local organizations. This has become one of the value additions of SEJAJAR network during March–June 2020. It also facilitated the spread and transmission of messages from government to local civil society and lower-level governments.

The case from NTT province resembles that of a networked CSO that mobilized to 'fix' the government's response system. Li et al. (2020) reported a similar gesture of 'fixing the state' whereby SEJAJAR's responses to the COVID-19 crisis fill the gaps in the formal response system led by both central and local governments in Indonesia. As a knee-jerk response to COVID-19 pandemic, SEJAJAR and FAN networks added unique value proposition to the whole of civil society approach to deal with COVID-19. Both CSO networks has performed multiple roles all at once: From a legitimate information clearing house; a knowledge broker; a platform for knowledge sharing; and a provider of technical assistance for sub-national NGOs; and a catalysator of government-civil society coordination.

References

Arief A (2020) Krisis Data dalam Penanganan Pagebluk. https://bebas.kompas.id/baca/bebas-akses/2020/08/05/krisis-data-dalam-penanganan-pagebluk/. Akses terahir 10 Agustus 2020

BNPB (2008) Pedoman Pembentukan Badan Penanggulangan Bencana Daerah. Peraturan Kepala BNPB 3/2008, 11 November 2008

Campbell-Nelson K (2003) Learning resistance in west Timor. PhD thesis, University of Massachusetts Amherst

Danusiri A (2020) Hantu, Virus, dan Semiotika Ketakutan. https://news.detik.com/kolom/d-5006641/hantu-virus-dan-semiotika-ketakutan. Akses terakhir 8 July 2020

Djalante R et al (2020) Review and analysis of current responses to COVID-19 in Indonesia: period of January to March 2020. Prog Dis Sci. https://doi.org/10.1016/j.pdisas.2020.100091

Eburn M, Dovers S (2014) Legal aspects of risk management in Australia. J Integrated Dis Risk Manag (IDRiM) 4(1):61–72

Lassa J (2020) Respons Masyarakat Sipil Atas Norma 'Adaptasi Kebiasaan Baru'. In: Winanti PS, Mas'udi W (eds) New normal: Perubahan Sosial Ekonomi dan Politik Pasca COVID-19

Lassa J, Booth M (2020). Are populist leaders a liability during COVID-19? The conversation

Lassa et al (2018) Twenty years of community-based disaster risk reduction experience from a dryland village in Indonesia. Jàmbá J Disaster Risk Stud 10(1):a502. https://doi.org/10.4102/jamba.v10i1.502

Li DE, Lassa J, Talan J Vieleta M, Banunaek R (2020) SEJAJAR coordination mechanism in response to the COVID-19 pandemic in Indonesia. SEJAJAR and Humanitarian Advisory Group's Commissioned Report by IRGSC

Li DE (2020a) Seandainya para bupati dan walikota di NTT mau berpikir? Timor Express 22 July 2020. https://timexkupang.com/2020/07/22/seandainya-para-bupati-wali-kota-di-ntt-mau-berpikir/. Akses terakhir 8 July 2020

Li DE (2020b) Isu-Isu Kelembagaan di Masa Pandemic: Pengalaman Forum Academia NTT. Seminar Kelembagaan, Universitas Satya Wacana, 14 Agustus 2020

Magar MT (2019) Housing recovery and reconstrucion platform. Asia Pacific Regional Workshop 2019 21 August 2019, Lombok

Mietzner M (2020) Populist anti-scientism, religious polarisation, and institutionalised corruption: how Indonesia's democratic decline shaped its COVID-19 response. J Curr Southeast Asian Affairs. https://doi.org/10.1177/1868103420935561

Quiggin J (2007) The risk society: social democracy in an uncertain world. Centre for policy development occasional paper number 2. Centre for Policy Development, Sydney. https://apo.org.au/node/2675. Last accessed: 03 July 2020

Said AA (2019) Jakarta Raih Indeks Demokrasi Tertinggi di Indonesia: Selain Jakarta, empat provinsi lain yang meraih Indeks Demokrasi Indonesia di atas 80 adalah Bali, Nusa Tenggara Timur (NTT), Kalimantan Utara, dan DIY. https://katadata.co.id/hariwidowati/berita/5e9a50d6bdde7/jakarta-raih-indeks-demokrasi-tertinggi-di-indonesia. Akses terakhir Tanggal 10 Agustus 2020

Tan P (2020) Terorisme Negara di Besipae. https://nttprogresif.com/2020/08/10/terorisme-negara-di-besipae/. Akses terakhir Tanggal 10 Agustus 2020

WHO (2010) Monitoring the building blocks of health systems: a handbook of indicators and their measurement strategies. World Health Organization

Widaningrum A, Mas'udi W (2020) Dinamika Respons Pemerintah Nasional: Krisis Kebijakan Penanganan COVID-19. In: Mas'udi W, Winanti PS (eds) Tata Kelola Penanganan COVID-19 di Indonesia: Kajian Awal. Gadjah Mada Press (Chapter 3)

Epilogue

The collective insights presented throughout the book offer a compelling reflection on disaster risk reduction. The threads that weave together the social, cultural, and economic aspects of disaster and crisis management remind us that addressing risk requires far more than technical interventions. It demands a deep understanding of how communities function, what they value, and how they respond when faced with crises.

One of the central lessons that emerges from these pages is the power of social cohesion. Time and again, we see that when formal systems are strained or slow to respond, the strength of community networks keeps people and society afloat. Whether during a pandemic or a natural disaster, local relationships—grounded in trust, mutual support, and shared knowledge—are critical to resilience. These social bonds are not just informal safety nets but the foundation upon which effective disaster preparedness and recovery are built.

Another recurring theme is the need for disaster management strategies that resonate with local cultures. Disaster risk communication, for instance, is most effective when it acknowledges the cultural contexts in which it operates. Communities have their own ways of interpreting risk, often shaped by generations of lived experience. For risk communication to truly take hold, it must align with these local narratives rather than imposing external frameworks that feel disconnected from the reality on the ground.

This recognition of local knowledge extends to governance as well. Risk governance, to be truly inclusive and sustainable, must be rooted in the values and belief systems of the people it serves. Governance models that fail to consider these cultural dimensions may struggle to gain traction, no matter how well-designed they are from a technical standpoint. Integrating cultural values into governance is not just a matter of respect; it's a practical necessity for building trust and fostering community participation.

At the same time, this book reminds us that disaster recovery is as much about mental and emotional healing as it is about physical reconstruction. Disasters can leave deep psychological scars, and addressing these is critical for long-term recovery. The importance of psychosocial support—often overlooked in disaster planning—cannot be understated. It is a crucial part of helping individuals and communities rebuild their lives in the aftermath of tragedy.

Economic resilience also plays a key role in the broader tapestry of disaster risk reduction. Whether through innovative financial mechanisms or adaptive social protection systems, there are practical ways to cushion the economic blow that disasters can deliver, particularly to the most

D. Hidayati et al. (eds.), *Examining Disaster Risk Reduction in Indonesia*, Sustainable Development Goals Series, https://doi.org/10.1007/978-981-96-4182-6

vulnerable. These strategies not only mitigate immediate financial impacts but also contribute to building longer-term resilience, allowing communities to recover more quickly and sustainably.

The findings serve as a reminder of the power of local knowledge, community cohesion, and networked governance. In many ways, CSOs act as the connective tissue that holds society together when formal institutions are slow to respond. As crises become more frequent and complex in our interconnected world, the lessons from this study will prove invaluable for shaping more inclusive, adaptive, and effective disaster governance frameworks that recognize the indispensable role of civil society.

The reflections on CSO activities in Indonesia during the pandemic reaffirm the need for governments to engage more deeply and transparently with civil society, recognizing them as partners and critical actors in times of uncertainty. The future of disaster governance lies in the balance of power and responsibility between the state and civil society—a balance that can only be achieved through trust, mutual respect, and collaborative action.

And finally, the role of international cooperation in disaster response serves as a reminder of how interconnected our world is. In times of crisis, collaboration across borders—whether through humanitarian aid or diplomatic agreements—can make the difference between a swift, effective response, and prolonged suffering. Yet, such cooperation must be carefully navigated to ensure that it respects national sovereignty and the unique needs of local populations.

In essence, this book brings to light the complexity of disaster risk reduction and the many layers that must be considered when crafting policies and strategies. From social capital to governance, economic resilience to mental health, the multifaceted nature of disaster management requires an approach as dynamic as the challenges themselves. As we move forward in a world where disasters—both natural and man-made—are increasingly frequent and severe, the lessons contained here offer valuable guidance. They remind us that disaster risk reduction is, at its core, about people—about their stories, their strengths, and their capacity to adapt and thrive in the face of adversity.

MIX
Papier aus verantwortungsvollen Quellen
Paper from responsible sources
FSC® C105338

If you have any concerns about our products,
you can contact us on
ProductSafety@springernature.com

In case Publisher is established outside the EU,
the EU authorized representative is:
Springer Nature Customer Service Center GmbH
Europaplatz 3, 69115 Heidelberg, Germany

Printed by Libri Plureos GmbH
in Hamburg, Germany